SUSTAINING T

Chris Reij is a geographer at the Centre for Development Cooperation Services, Free University of Amsterdam who has worked in many parts of Africa and Asia on soil conservation.

Ian Scoones is an agricultural ecologist with experience largely from Southern Africa. He is currently a Fellow of the Institute of Development Studies, University of Sussex.

Camilla Toulmin has worked mainly in Sahelian West Africa on agro-pastoral systems, land tenure and institutional issues. She is Director of the Drylands Programme at the International Institute for Environment and Development.

SUSTAINING THE SOIL

Indigenous Soil and Water Conservation in Africa

Edited by Chris Reij, Ian Scoones
and Camilla Toulmin

from Routledge

Photo credits:

Plates 1, 2, 4, 5, 6, 7, 8, 9, 10, 11, 13, 22, 23 and cover photo by Chris Reij
Plate 3 by Abdellah Laouina
Plate 14 by Wim Spaan
Plate 15 by Michael Mortimore
Plates 16 and 17 by Jürgen Hagmann
Plate 18 by Patrick Sikana
Plates 19, 20 and 21 by Anderson Lema
Plates 24, 26 and 27 by Hans-Joachim Kruger
Plate 25 by Million Alemayehu
Plates 28, 29 and 30 by Kevin Phillips-Howard
Plates 31 and 32 by Paul Tchawa

First published by Earthscan in the UK and USA in 1996

For a full list of publications please contact:

Earthscan
2 Park Square, Milton Park, Abingdon, Oxon OX14 4RN
Simultaneously published in the USA and Canada by Earthscan
711 Third Avenue, New York, NY 10017

Earthscan is an imprint of the Taylor & Francis Group, an informa business

A catalogue record for this book is available from the British Library

ISBN 978-1-85383-372-4 (pbk)

Typesetting by DP Photosetting, Aylesbury, Bucks
Cover design by Gary Inwood

CONTENTS

v

CONTENTS

FOREWORD

This book is about the importance of traditional soil and water conservation (SWC) practices in Africa. It clearly shows that many traditional conservation practices continue to be maintained and expanded, whereas modern SWC facilities are often poorly constructed and not adequately maintained. In recent decades, development agencies have invested a lot of resources in SWC measures in Africa and elsewhere with the aim of halting environmental degradation. But many of these efforts have had poor results, with land degradation persisting in many areas. This book provides a new perspective on these dilemmas and offers some challenging insights and new ways forward. The research on which it is based was funded by the Environment Programme of the Directorate General for International Cooperation of the Dutch Ministry of Foreign Affairs.

This book is not the result of western experts expounding their views on what should be done in this field in Africa. On the contrary, it is the result of research done by 30 teams of African researchers working for universities, ministries and NGOs, rediscovering the traditional knowledge of farmers throughout the continent. In their efforts they have been supported by the Centre for Development Cooperation Services (CDCS) of the Vrije Universiteit, Amsterdam and the London-based International Institute for Environment and Development.

The many case studies provide a number of important technical messages, but they also allow us to draw some policy lessons. Firstly, outsiders often assume that African agriculture is stagnating and unable to change. This view is largely prompted by catastrophic food shortages in a number of African countries coping with drought and war or a combination of both. This book helps to correct this general picture as it shows clearly that in many situations African farmers are able to feed themselves by continuously innovating and adapting to new circumstances.

Secondly, it shows that traditional soil and water conservation techniques can make a significant contribution to sustainable development in Africa. Traditional soil and water conservation techniques have long been ignored or ridiculed, but it is evident that they have a role to play. What I like is that this book does not take too radical a swing; it does not blindly glorify traditional soil and water conservation techniques, but makes a plea for marrying traditional and modern knowledge whenever possible.

Thirdly, most of the case studies in this book are from arid and semi-arid areas of Africa. In some cases (Niger, Burkina Faso and Mali) farmers have obtained spectacular results with the rehabilitation of badly degraded land. In Niger and Burkina Faso in particular, farmers in specific regions have spontaneously adopted improved traditional conservation methods on a very large scale. This shows that the implementation of the UN Convention to Combat Desertification, to which I attach a great deal of importance, does not need to start from scratch, but can be rooted in successful local experience. Again, this type of evidence counterbalances the usual gloom and doom scenarios for dryland Africa.

Fourthly, we have for a long time ignored indigenous knowledge, not only in soil and water conservation, but in agriculture in general as well as in the health sector. I believe that a more consistent effort to study and make use of indigenous knowledge systems is essential. At the same time this reassertion of the role of African indigenous knowledge is important in a cultural sense, because it demystifies the role of modern science as well as of western experts and it makes Africans feel proud of their own cultural heritage.

Finally, even when the macro-economic and policy conditions are not favourable for agricultural change, a growing number of farmers seem to adopt simple and cost-effective soil and water conservation technologies. They do so in particular in situations where they are facing serious environmental degradation and where their options are limited. Either they invest in improved land management, or they are forced to migrate. A major challenge now is to create a macro-economic and policy framework which induces land users to invest in improved resource management practices. This is one of the major objectives of the UN Convention to Combat Desertification and an important reason to support activities under this Convention.

This book should be of interest to all policy makers, researchers and fieldworkers involved with the support of sustainable development in Africa and elsewhere.

Jan Pronk,
The Hague,
15 July 1996

ACKNOWLEDGEMENTS

This book is the result of the enthusiasm of nearly one hundred African researchers who developed the 27 case studies of indigenous soil and water conservation (ISWC) in the 14 countries presented in this book. The original objective was to carry out up to 20 case studies but, due to the large number of positive reactions from African researchers, it was decided to increase the number. Without the dedication, interest and hard work of the African researchers involved, this project would not have been successful.

Funds were generously provided by the Environment Programme of the Directorate General for International Cooperation of the Ministry of Foreign Affairs, the Netherlands. We are particularly grateful for the unfailing support of Henri Jorritsma and Marjolijn van Deelen for this project and for their understanding of the complications of managing a project on this scale.

We would also like to thank Kebede Tato, Hans-Joachim Krueger and the team of the Soil Conservation Research Project for the organisation of a workshop in Addis Ababa in June 1994 which provided an opportunity to exchange many ideas and experiences at an early stage of the fieldwork. It proved to be one of the highlights of the project. A special thanks is due to Ben Haagsma, who was a member of the support group until February 1, 1995. We very much missed his management input into the project.

Administrative support was ably provided by Lia de Groot (CDCS) and Nicole Kenton (IIED). Michiel Knoop (CDCS) had the difficult task of transferring research funds to all corners of Africa, which often turned out to be very complicated. French case studies were translated into English by Jean Lubbock, Sarah Aljabar and Ann Withers. Editorial assistance was provided by Cathy Green, Dylan Hendricksen and Barbara Morris. The excellent maps for each case study area were created by Alie van de Wal (CDCS), and diagrams were redrawn where necessary by Helena Yeaman.

Finally, it was with great sorrow that we heard of the sudden death of Dr Kevin Phillips-Howard in a car crash in October 1995. As Head of the Department of Geography in the University of Transkei, he had been a most enthusiastic participant in the project and was involved closely with two case studies. His kindness and dedication will be remembered by all who knew him. We also have heard in the last few weeks of the death of Anderson Lema from Tanzania, who was a very lively and willing collaborator on this and previous research projects. His contributions to future research and scholarship in Tanzania will be greatly missed.

Chris Reij
Center for Development Cooperation Services, Vrije Universiteit, Amsterdam
Ian Scoones
Institute of Development Studies, University of Sussex, Brighton
Camilla Toulmin
International Institute for Environment and Development, London
July 1996

1

SUSTAINING THE SOIL

Indigenous soil and water conservation in Africa

Ian Scoones, Chris Reij and Camilla Toulmin

INTRODUCTION

Soil erosion is widely perceived to be a major problem in sub-Saharan Africa. Most agency reports and government publications highlight the degradation of soils as a major development challenge, but soil and water conservation (SWC) efforts in Africa have had a chequered history. From the early colonial era to the present, attempts have been made to introduce SWC measures in a wide range of settings, yet many have failed. This overview chapter asks why this is so, and whether there is an alternative approach which builds on local traditions in soil and water management. Drawing on the wide range of case studies presented in this volume, this overview asks the following questions:

- What are the key characteristics of locally managed SWC systems?
- What are the conditions for their successful adoption and expansion?
- How can development approaches be more effective in promoting the process of local-level SWC technology development?

SOIL EROSION IN AFRICA: MYTHS AND REALITIES

One of the overriding assumptions driving current policy and informing development interventions across Africa is that soil erosion constitutes a major problem. Global strategies such as the Desertification Convention, national environmental action plans and project documents are replete with statements that exhort us to combat soil erosion (Box 1.1).

BOX 1.1: THE POWER OF NUMBERS, THE NEED FOR ACTION: SOME SELECTED QUOTES

'A staggering total of 87% of the Near East and Africa north of the equator are in the grip of accelerated erosion' (FAO, 1983, p 4)

'Each year, 75 billion metric tons of soil are removed from the land by wind and water erosion' (Pimentel et al, 1995, p 1117)

'Fifteen percent of the Earth's total land surface is affected by human-induced processes of soil degradation... An estimated 6–7 million hectares of agricultural land is made unproductive each year because of erosion... Land degradation is widespread in the world's drylands, affecting 5.5 million hectares or almost 70% of their area and leading to an estimated loss of production worth US$42 billion' (IUCN/UNEP/WWF, 1991, p 110)

'Desertification affects about one sixth of the world's population, 70 per cent of all drylands, amounting to 3.6 billion hectares, and one quarter of the total land area of the world' (UN, 1992)

'About 5 to 10 million hectares annually become unusable due to severe degradation. If this trend continues, 1.4 to 2.8% of existing agriculture, pasture and forest land will be lost by 2020' (IFPRI, 1995, pp 10–11)

Such statements are based on a variety of sources, ranging from 'informed guesswork' to scientific measurement. The most influential are the studies carried out over the last 50 years which measure the degree of soil loss resulting from different agricultural and livestock management practices. Most of these studies derive from plot-based measurements which are then extrapolated to estimate total soil loss per hectare. There is no problem with such measurements when the measurements and the scale they refer to are fully acknowledged, but too often this is ignored and figures are extrapolated from a small plot to wider and wider scales. Conclusions drawn from such extrapolations are largely meaningless (Stocking, 1993). As an increasing number of studies now show, the soil loss estimates produced simply do not match levels of river or dam siltation in the same catchment (Walling, 1988). But the 'lost' soil must be going somewhere. Not surprisingly, studies show that much soil is redistributed within the agricultural landscape rather than being lost permanently (Bojo and Cassells, 1995). Box 1.2 provides some examples of influential studies where data derived from plot estimates at a field level have been extrapolated to national or even global levels, sometimes with dollar cost estimates attached.

Dramatic figures offer a powerful message to policy-makers: disaster is imminent and something must be done immediately (Blaikie, 1985). All major soil conservation efforts this century have been justified in these terms, from the colonial campaigns to introduce contour ridging in Southern Rhodesia (now Zimbabwe) and Kenya, to the large-scale mechanical conservation works in Burkina Faso in the 1960s and in Niger since the mid-1980s, to the massive food-for-work supported projects in the Ethiopian highlands in the 1980s.

BOX 1.2: SOIL LOSS DATA: FACTS AND FICTIONS?

Based on trials carried out in the 1950s (Hudson, 1957), estimates of field erosion in the communal areas of Zimbabwe have been put at 50–75 t/ha/year (Elwell, 1985). The financial cost of nutrient loss through soil erosion has been calculated at Z$2.5 billion (at 1985 prices) for the whole country (Stocking, 1986).

Assuming soil loss rates in Mali similar to those found on experimental plots in neighbouring Burkina Faso, the estimated capitalized net revenues forgone were calculated at between $7 and $26 per hectare per year, depending on the sensitivity of different crops to soil loss (Bishop and Allen, 1989).

Annual soil loss in Ethiopia is estimated at between 1.5 and 3 billion tonnes. Of this, 50% occurs in crop lands where soil loss may be as high as 296 tonnes/ha/year on steep slopes (FAO, 1986; SCRP, 1987).

In South Africa 300–400 million tonnes of soil are lost annually, equivalent to 10 tonnes per capita per annum (Huntley et al, 1989).

If on-site and off-site costs are combined, the total cost of erosion from agriculture in the United States is about US$44 billion per year (Pimentel et al, 1995, p 1121).

Of the 75×10^9 tons of soil eroded worldwide each year, about two thirds comes from agricultural land . . . this massive loss of soil costs the world about US$400 billion per year, or more than $70 per person per year (Pimentel et al, 1995, p 1121).

This perceived imperative for urgent action often provides great impetus to development debates. Nothing is more effective than the image of a dramatic gully to get people moving. A doomsday scenario gives out a powerful policy message. The rhetoric of Malthusian apocalypse, precipitated by environmental collapse, has been an enormously powerful force driving the agenda for external intervention. Such 'development narratives' (Roe, 1991, 1995) provide a simple, and apparently convincing story which leads to action. As Allan Hoben (1995, p 1008) explains:

> The environmental policies promoted by colonial regimes and later by donors in Africa rest on historically grounded, culturally constructed paradigms that at once describe a problem and prescribe its solution. Many of them are rooted in a narrative that tells us how things were in an earlier time when people lived in harmony with nature, how human agency has altered that harmony, and of the calamities that will plague people and nature if dramatic action is not taken soon.

The existence of 'crisis narratives' may have political implications, whereby external intervention and taking control over resources is justified because 'something just has to be done'. Emery Roe (1995, p 1066) argues that:

> Crisis narratives are the primary means whereby development experts and the institutions for which they work claim the rights to stewardship over land and resources they do not own. By generating and appealing to crisis narratives, technical experts and managers assert rights as 'stakeholders' in the land and resources they say are under crisis.

3

While it is certainly true that soil erosion is undermining agricultural production in certain places, the crisis may not be as prevalent as some commentators suggest. Indeed, our concentration on soil loss as *the* major issue has eclipsed other important production constraints, such as the water quality available to plants, soil nutrient levels, labour availability, market incentives, and so on. Rather than universal proclamations, a more sophisticated debate on the nature of soil erosion, its implications and consequences, is urgently needed (Stocking, 1996).

STANDARD SOLUTIONS FOR STANDARD PROBLEMS: THE CONVENTIONAL APPROACH TO SWC

Alarm about the potentially damaging consequences of soil erosion has prompted a long history of external intervention in SWC measures in Africa, as elsewhere (Hudson, 1987, 1991; Hurni and Kebede Tato, 1992; Pretty and Shah, 1994). The experience of the Dust Bowl in the United States proved highly influential in policy thinking from the 1930s onwards (Anderson, 1984; Beinart, 1984). This was compounded in southern Africa by the experience of drought. In South Africa the Drought Commission reported in apocalyptic tones:

> Enormous tracts of the country have been entirely or partly denuded of their original vegetation, with the result that rivers, vleis and water holes described by old travellers have dried up or disappeared ... the logical outcome of it all is the Great South African desert, uninhabitable by man (Anon, 1925, p 771).

The prospect of such disasters afflicting the newly established colonies worried many administrators and politicians. Some drew parallels with the collapse of previous civilizations, and a number of alarmist articles appeared, such as Lowdermilk's (1935) piece on 'soil erosion and civilisation'. In the British colonies agricultural engineers were quickly deployed to seek solutions. North American and European expertise was applied to the problem, along with some experience from work in India (Grove, 1995). The result was the emergence of a set of interventions focused on the mechanical conservation of soil: soil bunds, ridging, contour ploughing, and so on. In some parts of Africa, where the colonial state was strong and there was a need to ensure that demands for land by African farmers did not undermine the expansion of large-scale, European-owned commercial farming enterprises, major programmes of soil conservation were initiated. For instance, in the Machakos district of Kenya a forced programme of terracing during the 1950s resulted in the building of around 5000km of new terrace each year (Tiffen *et al*, 1994; Mortimore and Tiffen, 1995). A similar pattern occurred in much of southern Africa where major targets for contour bunding were set (Showers, 1989). In Southern Rhodesia (now Zimbabwe), the colonial authorities enforced the building of over 7000km of bunds in the period between 1929 and 1938. This continued through the following decades and by 1957 over 200,000ha of communal land had been treated (Whitlow, 1988). Similar policies were followed in Malawi where 118,000km of bunds were built between 1945 and 1960 (Stocking, 1985) and in Zambia where contour ridges were widely constructed in the eastern Province in the 1940s and 1950s (Mukanda and Mwiinga, 1993). By contrast, in the French colonies of West Africa no major soil conservation programmes were undertaken, with the exception of a pilot project in Burkina Faso in the late 1950s (Reij, 1983).

By the late 1940s, a wider set of environmental concerns had come to influence colonial development thinking: soil fertility decline, overgrazing and deforestation had been added to the list of ills inflicted on the land by African farming and livestock husbandry. This more comprehensive view of the environmental problem suggested a wider approach to land management which went beyond individual SWC techniques. By the 1950s an era of land-use planning emerged based on a set of land husbandry principles. In Kenya, the Swynnerton plan set out an ambitious vision for transforming the degraded environment of the Kenyan countryside. The Native Land Husbandry Act in Southern Rhodesia and the 'Betterment' Schemes in South Africa offered a strikingly similar set of solutions, involving the reorganization of land use according to conservation principles, the building of contour bunds linked to field drainage systems and the adoption of a mixed farming model.

In many areas the land husbandry package was rejected by local people. Farmers felt that the imposition of a particular model of land use practice undermined their existing agricultural management practices. In Southern Rhodesia colonial policies banned traditional wetland and river-bank cultivation, thus limiting people's coping strategies in dry years; they enforced reductions in cattle numbers, undermining people's ability to survive during drought; and they forced people to build a standard design of contour ridge to conserve soil and drain away water from the field, often with detrimental effects on productivity (Scoones et al, 1996; Wilson, 1988). Not surprisingly, in such situations colonial soil conservation and land husbandry measures were resisted, and in many countries they became the focus for nationalist opposition in the rural areas, leading to the widespread destruction of conservation structures, as a form of political protest.

However, in some places the suite of techniques and management practices offered by the land husbandry approach was widely adopted by farmers eager to invest in agricultural transformation. The well-documented case of parts of Machakos district in Kenya is a good example. Here, increasing population densities and resultant land scarcity combined with improved access to the growing market of Nairobi. Access to information through informal networks, as well as formal extension advice, enabled farmers to try out a range of conservation measures. From the mid-1960s, a major transformation of the farming landscape took place with huge voluntary investment in conservation works resulting in falling erosion rates, increased environmental rehabilitation and a boost in agricultural productivity (Tiffen et al, 1994). Another example of a successful transformation can be found in Swaziland. About 112,000km of grass strips were laid out on cultivated land between 1949 and 1960. This was achieved only because the Swazi king obliged all farmers to introduce grass strips on their fields through an Order-of-the-King. Despite attempts by USAID-funded soil conservation projects in the 1970s to discredit grass strips and to remove them, they remain a prominent feature in Swazi agriculture (Reij and Osunade, this volume).

During the 1960s and 1970s, development attention focused on the need to modernize and transform supposedly backward agriculture to raise yields and productivity. SWC technologies, it was argued, could play their part, and some grand schemes were dreamt up. Development planners who had visited the southern United States or the Negev in Israel had grand plans to transform the drylands of Africa into green oases. Large development budgets combined with access to sophisticated engineering designs and hardware ranging from theodolites to bulldozers. It seemed that anything was possible. Large project areas across Africa became the experimental laboratory for ambitious engineers with a vision. The results of these adventures are now well known. For instance, in northern Nigeria, attempts to encourage irrigated agriculture around Lake Chad collapsed very quickly with changes in the environment (Kolawole, this volume). Equally, in Burkina Faso, the 'Groupement Européen pour la Restauration des Sols' (GERES) project used earth-moving machinery to construct a dense network of bunds. However, only two-and-a-half years after the project's start in 1962, it was abandoned as farmers were not willing to maintain the bunds. Engineering solutions did not bring about a sudden modernization of agriculture; other constraints lay in the way of dramatic transformation. An African Green Revolution, along the lines occurring from the 1960s in the rice-growing areas of Asia, was much more elusive.

But developers were not to be put off, for the 1970s provided another spur for action. Much as the 1920s drought in southern Africa had sent shock waves through the British colonial administration in Africa, the 1972–73 drought in the Sahel set alarm-bells ringing in the international development community. Some of the most emotive media images in recent times have been of destitute and starving people, struggling to survive in apparently barren, degraded landscapes of dry Africa. These have been splashed across newspapers and transmitted through television screens throughout the world. This period coincided with emerging public interest in environmental issues in Europe and North America. The oil crisis of the early 1970s prompted a renewed wave of environmentalist concern. The Club of Rome's *Limits to Growth* (Meadows *et al*, 1972), Paul Ehrlich's *Population Bomb* (1968) and *The Ecologist's Blueprint for Survival* (1971) all helped to set the parameters for a debate premised on a possible collapse of the global ecosystem. Futurologists predicted an environmental doomsday, a concept which was only reinforced by petrol queues in Europe and pictures of starvation coming from Africa (North, 1995).

It was during this period that the term 'desertification' entered the international development lexicon. It had been widely used first many years before by Aubreville during the period of environmental furore that had characterized the 1930s and 1940s (eg, Aubreville, 1949). However, with growing concern about the links between environmental degradation and famine during the 1970s, the term was revived. At the policy level, a growing swell of concern culminated in the United Nations Conference on Desertification of 1977 held in Nairobi (UNCOD, 1977). Here, delegates from around the world committed themselves to a global plan to combat desertification. The militaristic imagery, captured in the language of 'combat', 'defeat', 'task-force', and so on, which characterized many of the official documents, only helped to re-emphasize the urgency of the task. Statistics galore were produced to illustrate the scale of the problem. In a bizarre exercise orchestrated by the United Nations Environment Programme (UNEP), questionnaires were sent to ministry officials in all countries asking for estimates of the areas affected by different degrees of desertification according to a scale ranging from very severe to none (UNEP, 1984). This 'data' was then projected on to official maps and simple sound-bite statistics were generated to illustrate the severity of 'the problem' (Oldeman et al, 1990; Box 1.1). It is illustrative of the political interests involved that, despite the clear fallibility of the data used, such information was raised to the status of fact and formed the basis for major international investments in environmental protection in Africa over the following decade (Swift, 1996). Thus, during the 1970s and 1980s, 'anti-desertification' projects were commonplace in Africa, and SWC measures were central to their design.

Many of these projects were remarkably reminiscent of the earlier large-scale colonial interventions. For instance, in Ethiopia, following the devastating drought of the mid-1980s, the Derg regime, backed by international aid funds and food relief, initiated a programme of SWC in the highlands. The influential Highlands Reclamation Study (FAO, 1986) provided the justification and the familiar range of soil conservation measures provided the technological solutions. The study predicted that by the year 2010, 7 per cent of the highlands would be bare rock, 11 per cent would have soil depths of less than 10cm and a further 76,000km^2 would be incapable of sustaining agriculture. These were dramatic conclusions. A standard package of interventions was evolved (Hurni, 1986) and implemented on a wide scale throughout the latter half of the 1980s. The Derg regime's political need to exert control over the rural areas coincided with the flood of food aid and development projects that arrived in the aftermath of the 1984 drought. Around US$20 million was disbursed annually as food-for-work between 1980 and 1990 (Cheatle, 1993, p 224). At one level, the result was impressive. Thousands of kilometres of bunds were built, thousands of hillsides were closed off, steep-slope agriculture was abandoned and millions of trees were planted (IUCN, 1990). However, there were high costs. The lack of involvement of people in the planning and implementation of the schemes meant that they were often poorly executed and maintained. In addition, in some places the imposed measures disrupted existing SWC measures, replacing them with alternatives that were less suited to the local setting. The focus on soil erosion as the core problem also detracted attention from the wider problems faced by rural people; without addressing such issues, there was little chance that the soil conservation measures would be widely adopted without significant subsidy and, in some instances, coercion (Dessalegn, 1994).

At varying scales and to varying degrees, a similar story has been repeated in many parts of Africa (Hudson, 1991). By the late 1980s, a growing realization emerged that soil conservation was not the whole answer. Again, the parallels with the colonial period are striking. Indeed, the same terms were resurrected. Instead of soil conservation, a wider concept of 'land husbandry' was expounded (cf Hudson and Cheatle, 1993; Hudson, 1992; Shaxson *et al*, 1989). Such approaches argued for bringing together a wide range of technologies to deal with the broader developmental problems faced by smallholder farmers.

PUTTING PEOPLE FIRST: PARTICIPATORY APPROACHES TO SOIL AND WATER CONSERVATION

The one important difference between the advocacy of SWC and land husbandry today and colonial precursors is the current emphasis on people's participation. The lessons from the 1960s onwards taught project planners and policy-makers alike that imposed projects just do not work, certainly in the longer term. Advocates of a more participatory approach to development argued forcefully for 'putting people first' (Chambers, 1983). Wider trends of democratization, decentralization and the retreat of the state have meant that participation has become both politically appropriate and practically necessary.

This constellation of factors has prompted the emergence of a new style of natural resource management intervention that is based on holistic, village-based resource management involving a participatory process in planning and implementation. This is particularly well illustrated in the village land management (*gestion de terroirs*) approach which is widely promoted by donors in Sahelian West Africa (Box 1.3).

This type of approach is now widely endorsed and representative of a broad consensus about development intervention in the 1990s. Donors and national governments have adopted, at least at a rhetorical level, the language of participation and land husbandry. This is reflected at every level from village project plans, through national environmental strategies and plans to global conventions. The Earth Summit in Rio in 1992 successfully raised environmental concerns at the level of global politics for the first time since the early 1970s. The follow-up strategy for the next century contained in Agenda 21 is replete with hopeful statements about how people's participation is the appropriate route to effective land management (UN, 1992). The Convention on Desertification, in contrast to the global and national plans of action to combat desertification developed in the 1980s, is equally strong on the rhetoric of supporting local level, participatory processes, and placing less emphasis on seeking technological solutions to perceived environmental decline (Toulmin, 1994).

The shift to local-level planning, appreciation of indigenous techniques and

BOX 1.3: *GESTION DE TERROIRS*: A WAY FORWARD OR A BLIND ALLEY?

The last 10 years in the Sahel have witnessed an explosion in the number of projects calling themselves 'gestion de terroirs' (GT), which roughly translates as 'management of village lands'. In theory, the GT approach involves a series of steps from:

- a participatory diagnosis by local people to identify problems of resource degradation
- election of a committee at village level responsible for resource management decisions
- establishment of boundaries to the village territory
- preparation of a management plan, specifying the different zones and uses to which they will be put
- implementation, monitoring and evaluation of activities

The approach is appealing, based on a commitment to local decision-making and diagnosis of problems to be tackled. In practice, GT projects have run into difficulties such as: how to include mobile resource users (eg pastoral herders) into decision-making structures: the heterogeneous nature of rural society – poor and rich, women and men, old settler families and new migrants, each with different interests and demands; and the very limited real transfer by governments to local communities of power to control access to land, water and vegetation. Doubt has also been cast upon the usefulness of setting up new village level structures for resource management, when existing social institutions can do the job better.

The GT approach represents a much more promising way forward than previous attempts to promote rural development, particularly in the very dry, and risk-prone Sahel. But such promise will not be attained unless governments shift real decision-making power to their populations.

Sources: Toulmin, 1993; Painter, 1993.

acceptance that there are limits to technological solutions to complex land-management problems are undoubtedly a step in the right direction. But as experience emerges from this new generation of projects, some important questions are emerging. First, the distance between the rhetoric and reality of participation means that many projects are simply a new vehicle for the imposition of technological solutions. This time they may be more people friendly (agroforestry, woodland management, small-scale soil conservation and water harvesting systems, etc) than the grand engineering schemes of the 1960s and 1970s, but it is questionable whether they are any more sustainable once the project subsidies have been removed. Secondly, the well-polished populist rhetoric of community participation may mask hidden conflicts, diverse interests and unnoticed costs. For instance, the now well-known Yatenga project from Burkina Faso (Ouedraogo *et al*, this volume) has been extraordinarily successful at one level: stone lines have been built across huge areas with and without project support. Degraded areas have been rehabilitated; production has been boosted; village committees have been formed to manage the process; and the level of external subsidy has been reduced. But behind this apparent success lie other stories. Benefits and costs have been unevenly distributed, with élites capturing many of the rewards and women's labour costs increasingly significantly (Atampugre, 1993; Gubbels, 1994; David, 1995). Peter Gubbels (1994, p 241) observes:

> 'Putting farmers first' is striking, resonant rhetoric, but not easy to put into practice. It requires deciding *which* category of farmers should come first. *Not* deciding inevitably means that local élites come first. Indeed, to achieve goals such as promoting self-reliance, peasant organization and community environmental management, outside intervention is often not able to avoid working with rural power structures and may have to compromise on equity issues.

Characteristics of indigenous SWC techniques

Before describing some of the key characteristics of local SWC practice, it is important to establish what we mean by 'indigenous SWC techniques'. There has been a lot of debate in recent years about the importance of what has come to be known as 'indigenous technical knowledge' (Richards, 1985; Warren, 1991) in the process of technology development. But what is 'indigenous' and what is 'technology'? Both are contested terms and difficult to define. For our purposes, indigenous refers to local practices, as distinct from interventions imposed from outside. However, many practices that may be regarded as indigenous today may have been derived from elsewhere in the past. For instance, the stone bunding of the Harerge highlands of Ethiopia possibly had its origins in Arabia and was imported by traders coming to the area during the Middle Ages (Asrat *et al*, this volume). Similarly, the history of SWC measures employed in the lowland rice-fields of the Shinyanga region, Tanzania, can be traced to interaction between local people and Asian migrants working in the area during the 1950s (Shaka *et al*, this volume). Indeed, many of the 'indigenous' techniques described in the chapters that follow have their origins elsewhere, derived from migrants living in or passing through the area, learned during journeys to other places or adapted from interventions imposed during the colonial era. In this volume we are not concerned with a static notion of indigenous knowledge and technology, apparently frozen in time, stuck in history. Instead, we are interested in the dynamics of technical change, how innovations are adopted and transformed, how technologies evolve through incremental adaptation and how current practices are the result of cumulative responses to a range of influences over time.

This dynamic interpretation of indigenous SWC leads us to a wide-ranging perspective on technology. SWC technologies are not simply structures defined strictly by engineering parameters; they are the sum of practices involved in managing soil and water in agricultural settings and they also include agroforestry, agronomic and tillage practices (Reij, 1991). Any analysis of technology must therefore be situated within a social and economic understanding of the role of the technology, the rationale and purpose of its design. Technologies arise out of particular sets of historical and social circumstances, different people have different attitudes and commitments to them and, because of the dynamic influences over their origin and maintenance, they evolve and change continuously. For instance, in Zimbabwe a standard design of contour ridge was imposed during the colonial era. This was designed according to the highest engineering standards with the purpose of conserving soil through diverting water away from the field to a waterway. But this technology was inappropriate to the needs of farmers in the drier areas so, over time, contour ridges have been adapted from soil conservation to a water-harvesting function with the adoption of double-ridging techniques and the digging of moisture conservation pits (Hagmann and Murwirwa, this volume). This has caused outrage among some engineers who see this as a disruption of the engineering design. But to farmers the contour ridges now serve a useful purpose, whereas before they did not.

Thus, most local SWC practices have designs that reflect their multiple functions (Reij, 1991; IFAD, 1992). Soil conservation and water harvesting may have different priorities, depending on the average rainfall in the area, the soil types and the position of the site within the landscape. For instance, in wetter areas leaching of soil nutrients and sheet erosion may be a serious problem for agricultural production, and soil-conservation measures may be of paramount importance.

In the Bameleke highlands, Cameroon, for example, farmers invest in complex soil-management practices to improve production for their high-value crops (Tchawa, this volume). By contrast, in drier areas such as the Red Sea Hills or Darfur in the Sudan, water is the major constraint to agricultural production and so the technologies are designed to capture and spread water to key agricultural sites (Yagoub; El Sammani, this volume). This trade-off between soil-nutrient management and soil-moisture management is central to current understandings of the production dynamics of African savannah environments (Frost et al, 1986). The figure below illustrates

Agroecosystem trade-offs and the implications for SWC: examples from the case study sites

Available soil moisture →	
Dry/high nutrients: Water harvesting and spreading (eg. wadis in Sudan; flood plains in northern Nigeria)	Wet/high nutrients: Soil and water maintenance (eg. humid zone gardening in Cameroun)
Dry/low nutrients: Water and soil harvesting (eg. sandy soils in dry miombo areas of Zimbabwe and Tanzania)	Wet/low nutrients: Fertility management; soil harvesting (eg. Ethiopian highlands; northern Zambia)

(left axis, bottom to top: Available soil nutrients →)

this trade-off, showing how different SWC techniques are appropriate to different settings.

Technologies must also respond to environmental change. This may be through shifts in climatic patterns or local changes in land-form. For instance, in northern Zambia, changes in river flow and the creation of a lake opened up new livelihood opportunities, resulting in shifts in agricultural practices, new crops and with them new technical practices (Sikana and Mwambazi, this volume). Equally, in northern Nigeria the traditional *firki* cultivation system, combining dryland and wetland farming, is able to respond to wet and dry climate cycles and with them the expansion and contraction of the area of Lake Chad. This is in sharp contrast to introduced interventions which were designed on the assumption of stable, average conditions (Kolawole *et al*, this volume). In the Sahel, where water limitation is the major constraint, projects which concentrated on soil erosion control continued to be promoted throughout the 1980s, despite persistent drought. However, those techniques most rapidly adopted were water-harvesting techniques (Ouedraogo and Kaboré, this volume; Hassan, this volume).

Types of SWC technique also vary within landscapes. Generally, at the lower point in the slope, sink sites form where soil and water collect. Such areas include a variety of wetland patches within dry areas (Scoones, 1991), with examples including the *wadis* of Sudan, *mbuga* of Tanzania, *dambos* of Zimbabwe and Zambia, and lake-shore sites in Nigeria (see Chapters 2, 3, 11, 12, 13 and 15 in this volume). Within dry areas, such sites are highly valuable, constituting key resources in otherwise fairly low-value landscapes. The potential productivity of these low-lying areas is often high, although they tend to demand higher labour inputs than dryland cultivation. Under certain conditions, it may pay to invest in complex SWC measures in these areas in order to maintain high soil moisture and nutrient status, managing seasonal variations of flooding and drought in order to maximize production. For this reason, such sites tend to have the most elaborate forms of SWC technology. For instance, in Shinyanga, Tanzania, farmers have invested in elaborate bunding systems in their *mbuga* rice-fields, whereas their upland fields and grazing areas have little SWC investment associated with them (Shaka *et al*, this volume). Equally, when upland areas cease to provide the returns they once did because of declining rainfall or environmental degradation, bottom-land sites become increasingly attractive. In southern Tanzania, for instance, an elaborate form of raised-bed cultivation is increasingly being practised, with such sites providing income and food security for local farmers (Lema, this volume).

Unlike the conventional engineering designs of SWC structures which are specified in technical manuals and extension handbooks with precise dimensions and design requirements, local SWC techniques are much more flexible (Critchley *et al*, 1994). Flexibility is important as field topography varies greatly from site to site. Soil distribution within the field also changes, demanding new designs for new conditions. As illustrated by the case of northern Zambia, ethno-engineering is a result of 'adaptive performance', rather than a timeless response to a technical design problem (Sikana and Mwambazi, this volume; Richards, 1989). In Ethiopia, traditional stone bunds are constructed step by step and only get their final shape and dimensions gradually, over many years. Construction is thus attuned to levels of labour and materials availability. This offers farmers the chance to experiment and add modifications wherever necessary, rather than opting for a single design built in one go. Stone bunds can be removed and built up again at another location in the field; the spacing can change according to specific field conditions and a huge diversity of con-figurations are possible – different lengths, widths and heights can be found in any site. Thus, they are not permanent structures, and many remain at one position for perhaps five to ten years before they are shifted, as the form of the field slope, soil composition and drainage patterns change with the continuous process of erosion and deposition caused by agricultural use (Krüger et al, this volume).

Indigenous SWC measures therefore tend to spread labour requirements for construction and maintenance. Again, this contrasts with most introduced techniques which require major investments of labour in construction, often during a single period. Gender differentiation of labour inputs is common, with men often being involved in discrete, time-bound activities such as construction, while women are relied upon for on-going labour inputs such as the main-tenance of structures. This division may result in conflicts when men engage in construction (the most likely activity to be subsidized by development projects), while women are not prepared to invest their time in long-term maintenance. For instance, in northern Ghana this sexual division of labour is apparent, resulting in important implications for the way that SWC is perceived by women and men (Millar, this volume). Gender differences in labour investment may shift with changes in the value of different landscape components for farming. For instance, in southern Tanzania, men are increasingly taking up gardening in lowland sites, activities that were previously the preserve of women (Lema, this volume). This may result in changes in the way that gardening is practised, with new technologies with different labour requirements being adopted. However, in other areas, where high levels of male out-migration exist, such as in southern Malawi, women must take on the full range of tasks. In such areas, labour shortages may be an important reason for the low levels of investment in SWC on small holder farms (Mangisoni and Phiri, this volume). In the past, mobi-lizing group-based labour through co-operative work parties has been an important way of alleviating household-level labour shortages, especially for construction activities. However, a number of case studies report that such practices are on the decline as the social networks upon which they are based fragment (eg, Igbokwe; Mbegu, this volume).

Conditions for success

What are the conditions for the success of SWC systems? The case studies contained in this volume illustrate a wide diversity of situations, from areas with extremely low population density in the arid zones through to high population density areas in some semi-arid and humid zones, and from areas well served by infrastructure with good access to markets to remote areas far from urban centres, without good road connections and where subsistence production dominates. Equally, the histories of external intervention, access to information on new technologies and broader policy environments have all differed between the case study sites.

The table below summarizes some key characteristics of the case studies, including a description of the region and the ethnic or language group, contrasting rainfall, population density and crops grown and relating these to the SWC technique found at each site. The following chapters present each of these case studies, starting with those found in arid zones, moving on to semi-arid,

Case study characteristics: from drier to wetter areas

Country	Region	Rainfall (mm)	Population density persons/km^2	Ethnic/ language groups	Major crops	SWC techniques
Sudan (Ch. 2)	Red Sea Hills	25–150	1–10	Beja	sorghum, millet	earth bunds
Morocco (Ch. 4)	High Atlas Mountains	45–340	1–10	Mgouna	barley, wheat, maize	bench terraces
Sudan (Ch. 3)	Central Darfur	100–400	10	Fur, Zaghawa	sorghum	earth bunds
Morocco (Ch. 5)	Rif Mountains	350–450	10–100	Berber	wheat, barley, fruit trees, hemp	bench terraces, stone bunds, step terraces
Niger (Ch. 6)	Tahoua	350–450	22	Hausa	millet, sorghum	improved planting pits
Nigeria (Ch. 11)	Borno state	250–500	37	Kanuri	sorghum	earth bunds
Mali (Ch. 7)	Djenné	275–600	28	Bambara	millet, sorghum	improved planting pits
Mali (Ch. 8)	Dogon Plateau	500	25–80	Dogon	millet, sorghum, vegetables	micro-basins, pitting, stone bunds
Zimbabwe (Ch. 12)	Masvingo Province	400–600	45–60	Shona	maize, millet, sorghum	modification of contour ridges
Burkina Faso (Ch. 9)	Yatenga	400–700	20–130	Mossi	millet, sorghum	improved planting pits, stone bunds
Burkina Faso (Ch. 10)	Central Plateau	400–800	40–100+	Mossi	millet, sorghum	mulching, contour stone bunds

Case study characteristics: from drier to wetter areas (continued)

Country	Region	Rainfall (mm)	Population density persons/km²	Ethnic/language groups	Major crops	SWC techniques
Zambia (Ch. 13)	Northern Province	650–850	10+	varied	cassava, maize	raised-bed cultivation
Tanzania (Ch. 15)	Maswa District	600–900	50–70	Wasukuma	rice, cotton, maize, sweet potatoes	earth bunds
Ghana (Ch. 14)	Upper East	800–900	204	Frafra	sorghum, millet, groundnuts	stone bunding
Tanzania (Ch. 16)	Rukwa region	900–1000	30	Wafipa	maize, millet, beans	mounds
Cameroon (Ch. 24)	Mandara Mountains	400–1200	40–100	Mafa, Mandara	sorghum, millet	bench terraces, stone bunds
Malawi (Ch. 25)	Southern region	500–1300	220–292	Lomwe, Yao, Chewa, Sena	maize, sorghum	contour bunds, strips, vegetation barriers
Swaziland (Ch. 19)	Swazi Nation Land	400–1500	30–65	Swazi	maize	grass strips
Ethiopia (Ch. 20)	Harerge	700–1100	230–410	Oromo	sorghum, coffee, chat	earth bunds, bench terraces
Ethiopia (Ch. 21)	Northern Shewa	1350	70–200	Amhara	barley, wheat, pulses	drainage ditches
South Africa (Ch. 23)	Transkei, Eastern Cape	750–1400	84	Xhosa	maize	adaptation of contour banks
Tanzania (Ch. 17)	Njombe District	900–1600	30	Wabena	finger millet, maize, beans, potatoes	raised-bed cultivation
Tanzania (Ch. 18)	Mbinga District	900–2000	35–120	Matengo	coffee, maize, beans	pits
Nigeria (Ch. 26)	Jos Plateau	1000–1500	70–280	Berom, Hausa	vegetables, wheat, maize	basin irrigation
Nigeria (Ch. 27)	Enugu	1600–2000	335	Igbo	yams, maize, cocoyams, vegetables	bench terraces
Cameroon (Ch. 28)	Bamiléke Plateau	⩾3600	50–275	Bamileke	maize, cassava, yams	ridge cultivation, hedge barriers

15

sub-humid and humid areas. This gradient from dry to wet environments across Africa highlights some important differences in the context for SWC technologies. These include the relative importance of water versus soil harvesting and conservation; the inherent productivity of agriculture and so the incentives to invest in production; levels of population density and labour availability; the type of land tenure existing, and so on.

It is a difficult task to dissect the interaction of influences that condition the success or failure of particular SWC techniques in particular places at particular times. So many factors interact often in conjunction during periods of crisis that any simple explanation for technological development is always lacking without historical insights (see Chaker *et al*, Box 5.1; Slingerland and Masdewel, this volume). However, there are a number of important themes arising from the case studies, which are highlighted below.

Population densities

A wide range of population densities are found in the case study sites, ranging from less than 1 person per km^2 in the Red Sea Hills of Sudan to over 300 persons per km^2 in south-eastern Nigeria. As Ester Boserup (1965) and many others since (eg, Turner *et al*, 1993; Tiffen *et al*, 1994) have noted, population density has a major impact on the processes of agricultural intensification. With higher population densities, agricultural plot sizes shrink, and land, rather than labour, becomes a key constraint to production. This in turn provides incentives for investing in new technologies, conserving the resource base and through this, increasing production. For instance, in former Transkei, South Africa, a combination of high population density, owing to the impact of apartheid policies, and drought have forced people to invest in intensive garden management, including small-scale irrigation (Phillips-Howard and Oche, this volume). Equally, in Burkina Faso the combination of drought and high population densities has encouraged investment in *zaï* (planting pits) in the Yatenga region (Ouedraogo and Kaboré, this volume).

Access to land and labour, however, may be highly differentiated within a village community, resulting in different dynamics of intensification. Equally, within a household men and women may control different types of agricultural plots, resulting in different types of investment. Often women's plots are small and land-enhancing investments are most likely. For instance, in Zimbabwe women have access to small gardens, while men may have control over main-field sites. In this setting, women have experimented with a range of soil-and water-management practices for garden sites which require intensive management but provide high returns (Hagmann and Murwirwa, this volume).

Of course, a range of other factors must interact to encourage the process of intensification, but the evidence certainly suggests that rising population density is one of a number of important preconditions for investment in SWC measures. This argument runs counter to the oft-repeated Malthusian view that population growth will inevitably outstrip food supplies, resulting in environmental degradation, collapse in food production and ultimately starvation or forced migration.

What is the evidence to support these different scenarios from the case studies? It is certainly true that in the areas with the highest population density, we find the most elaborate SWC structures. For instance, in the Ethiopian highlands a huge range of SWC measures are found (Krüger *et al*, this volume). By contrast, the level of labour investment is far lower in the water-spreading devices found in low population density areas of the Sudan or northern Nigeria (El Sammani and Dabloub; Kolawole *et al*, this volume). In areas which have increasing populations, such as in the communal areas of Zimbabwe or in parts of Tanzania, a trend towards increased investment in SWC measures can be seen (Hagmann and Murwira; Lema, this volume).

By contrast, in those areas where out-migration has caused rural depopulation, particularly of males who were heavily engaged in construction, then decreasing levels of SWC investment are seen, as in the Mandara Mountains of Cameroon (Hiol Hiol, this volume) or in southern Malawi (Mangisoni and Phiri, this volume). However, in some areas, migrant labourers' earnings are used to hire labour to build and maintain SWC structures (Chaker *et al*; Box 5.1, this volume). In other areas, such as in the Atlas Mountains of Morocco (Hamza *et al*, this volume), SWC structures are maintained because the social value of migrants' rural land remains high, even if its productive importance has declined.

But population density provides only part of the explanation, for there are also cases with high population density coupled with low voluntary investment in SWC, such as in the ex-homelands of South Africa (Phillips-Howard and Oche, this volume), as well as areas with relatively low population densities where highly elaborate SWC measures are seen (eg, Yagoub; El Sammani and Dabloub this volume). The following sections explore a variety of other factors which are influential in providing the conditions for the successful spread of SWC.

Investment and access to capital

The lack of capital markets and, in particular, the lack of formal credit opportunities is often highlighted as a serious constraint to investment in new technology. However, the case studies presented in this volume do not highlight credit as a major constraint. The uncertain returns from investment in dryland agriculture, and conservation measures in particular, mean that formal credit arrangements are unlikely to work effectively. In any case, small-scale SWC does not require major capital investment and therefore there is a limited need to mobilize cash, except for the payment of labour in some cases. In much of dryland Africa, non-farm income is a key substitute for credit (Reardon *et al*, 1994). But whether such income is invested in agricultural technology depends on the profitability and riskiness of investment returns. The farmers in the Central Rif Mountains in northern Morocco invest in terraces because it allows them to grow a cash crop (marijuana or *kif*), which produces substantially higher returns per hectare than food crops. About 200,000 people in the Central Rif depend on *kif* and this region shows an acceleration in population growth rates (Chaker *et al*; Box 5.2, this volume). The growing popularity of a number of low-cost, traditional SWC techniques in the Sahel is not only due to the fact that they are simple, but also because they provide rapid and substantial returns to investment of labour (Hassan; Ouedraogo and Kaboré, this volume).

Returns to SWC investment

It is notoriously difficult to assess the returns to SWC investment, but some broad assessments have been made. A general conclusion from work carried out in Central America echoes findings from African settings:

> Except when high-value crops are planted on fragile soils, expensive mechanical structures are unlikely to be profitable... Conservation measures are particularly likely to be profitable either when they are cheap and simple, or when they allow farmers to adapt existing practices (Lutz *et al* 1994, p 289).

Studies on the returns to SWC in Africa are few and far between. Too often it is assumed that SWC is automatically beneficial, without looking in detail at the costs and benefits. Those few studies that have been carried out relate to large-scale mechanical conservation works and show that in most cases returns are negative. For instance, Herweg (1992) found that the recommended soil erosion measures for highland Ethiopia consistently provided negative returns under farmer conditions. Equally, Jan Bojo (1991) found that a major project involving farm improvement with SWC made a significant net loss relative to the resources invested, even when benefits to society were factored in. By contrast, those low-cost SWC measures designed by farmers may offer significant increases in production, especially when water-harvesting techniques promote yield increases and drought protection in semi-arid areas or when previously barren land is rehabilitated (Wedum *et al*, this volume). In Niger, the rehabilitation of previously degraded land has become so attractive that farmers and traders are buying land in order to rehabilitate it (Hassan, this volume).

Some proponents of SWC programmes argue that calculation of immediate returns should not be a concern because such measures are aimed at long-term conservation which must attract external subsidy in order to assure inter-generational equity and to offset the wider costs of erosion. But, given the constraints on government budgets and the contraction of aid flows to Africa, cost-effectiveness, even in the short term, will remain an important priority for project planners, as it always has for farmers. The limited occasions where subsidies may be justified would include instances where major divergences arise between private and social costs and benefits, such as cases where off-farm or downstream effects are significant (Lutz *et al*, 1994). Even in such cases, caution in offering subsidized inducements must be heeded.

Experience shows that subsidies may well alter behaviour and encourage investment, but this may not be sustainable and as soon as the subsidy is withdrawn, farmers switch to more cost-effective strategies. This pattern is highlighted again and again in the history of SWC in Africa where, following the removal of inducements, farmers dismantle SWC measures installed under previous regimes in favour of more appropriate, existing measures. For instance, in northern Nigeria following the collapse of the Lake Chad irrigation project, there was a rapid revival of *firki/masakwa* cultivation across a wide area (Kolawole *et al*, this volume). Equally, in Ethiopia, following the fall of the Derg regime and the shift in priority of food-for-work projects, many farmers have abandoned, or at least adapted, their SWC bunds which were implemented in the previous era. Similarly, in countries which have removed subsidies on inorganic fertilizers as part of Structural Adjustment Programmes, there has often been a widespread revival of organic fertility management practices (Mbegu, this volume).

Markets and infrastructure

The incentive to invest in intensification will increase as the value of the output rises. But, without good infrastructure and access to markets, the growth in economic incentives may not parallel demographic pressures and the spontaneous innovation and spread of SWC may not take place. However, if a good road system and competitively priced transport provide access to urban markets with high demand, then the crop values increase, resulting in higher incentives to invest for long-term gain.

The policy environment that governs both the working of markets and the patterns of public investment in rural infrastructure is also critical. In the past, state intervention in agricultural production and marketing has often constrained opportunities. For instance, in Tanzania, the low prices paid, the poor marketing arrangements and the limited service support provided, reduced people's willingness to grow cotton. By contrast, rice, which was unregulated by the state, offered a much better return and farmers switched to this crop, changing their agricultural production systems dramatically towards wetland cropping. This change required new technologies and new investments in SWC measures (Shaka *et al*, this volume). Structural adjustment policies in many African countries have resulted in the liberalization of markets, the abolition of parastatal marketing agencies and the encouragement of export crops through price reforms. This has had mixed effects on the incentive to invest in SWC measures. In some cases, such agricultural reforms have resulted in increased crop prices, increasing the incentive to remain in rural production. In other cases, reforms have meant that new crops are now favoured, resulting in changes to cropping patterns and new demands on SWC measures.

Two contrary views exist about the likely effects of market liberalization and improved crop prices on incentives to manage and conserve soils in Africa (Barrett, 1991; Barbier, 1991). Some observers expect that agricultural reforms which bring increased crop prices should encourage more people to remain in rural production and invest in the land. Hence, rising crop prices should bring higher levels of soil conservation, as farmers expect higher returns from their land now and in future (Repetto, 1988). Others argue that higher crop prices may just as well encourage a rapid soil 'mining', with farmers trying to gain the maximum immediate return from their soils, and no necessary increase in investment in the conservation of soils and their fertility (Lipton, 1987). The reaction of land users to crop price increases is likely to depend on several factors, such as their dependence on the continued cultivation of this land as a source of income. Where land is in plentiful supply, or where the cultivator can easily move into other fields of economic activity, there may be little long-term interest in maintaining soil fertility, so that rising crop prices bring accelerated mining of nutrients. For example, the Mouride brotherhoods in Senegal deplete soils very rapidly in their groundnut plantations, in large part because they know they can abandon the land and move on to newly cleared lands elsewhere. For small-scale farmers who are strongly embedded in a particular place, the option of moving elsewhere may not be realistic and, hence, they are much more likely to see a rise in crop prices as an incentive to invest further in soil conservation. The degree to which farmers will change their SWC practice as a result of price changes also depends on their willingness to trade-off income now against income in the future, and whether they expect confidently to benefit from current investments, itself dependent on security of tenure.

The liberalization of agricultural prices and currency devaluation have also led to changing cropping patterns, with exportable crops generating higher returns. There is no evidence to date, however, for export crops being associated with either more extractive or more conservative soil management. Structural adjustment and the liberalization of agricultural markets have also brought substantial increases in fertilizer prices as a result of devaluation and the abolition of subsidies. In general, inorganic fertilizer can be considered a substitute for labour invested in soil conservation and improvement. When fertilizer becomes more expensive, farmers are likely to economize on its use and to ensure that minimum losses occur by investing labour in soil management and conservation to prevent run-off of nutrients. Hence, it might be expected that investment in SWC will increase as inorganic fertilizer becomes scarcer. For instance, in Enugu state and the Jos Plateau in Nigeria, the onset of structural adjustment policies and the decline of oil revenues, resulting in reduced opportunities for off-farm work, have generated an increase in SWC and small-scale irrigation investment (Igbokwe; Phillips-Howard, this volume). The effects of liberalization are also influenced by levels of public investment to complement the reform policies. In places where no roads exist, where they are poorly maintained, or where limited transport results in high prices, any benefits of structural reform are barely felt. Indeed, with the withdrawal of subsidies on marketing (eg, through parastatal pricing), people living in such areas may suffer reduced incomes and lower incentives to invest in SWC.

Security and tenure rights

Investment in SWC will depend on the willingness of farmers to expend labour now for increased benefits which may be obtained in the first year (in particular in semi-arid regions) or some time in the future – for instance, in the case of agroforestry practices. This means that people must feel confident of secure benefits from this investment. But insecurity can arise in a variety of ways.

Conflict and war currently affect large parts of Africa, resulting in major disruption to rural life and production systems. Of the case study areas, there are ongoing conflicts in western Sudan and parts of central Africa, with major refugee movements to nearby countries or urban centres. Clearly, under such conditions people are unlikely to invest in SWC if they are unsure that they will be living at the same place in the next few months. However, war or raiding has sometimes resulted in the origin of major SWC investments, as people retreated into confined refuge areas (eg, Igbokwe for Nigeria; Kassogué *et al* for the Dogon Plateau in Mali).

Insecurity may also arise through heavy-handed development interventions. For instance, people expecting, or fearing, displacement may be unwilling to initiate SWC measures of their own. The history of SWC in Africa has unfortunately been characterized by forms of external intervention that have undermined local initiatives. Several of the case studies illustrate how, in the colonial era and since, externally planned measures have disrupted indigenous practices, often replacing SWC measures evolved by local people over many years. For instance, in Zimbabwe the imposition of a whole range of policies from the early part of this century resulted in the almost complete abandonment of indigenous practices (Hagmann and Murwirwa, this volume). Equally, in Ethiopia the expansion of soil-conservation measures, with the support of food-for-work programmes, has often resulted in the removal of indigenous measures and their replacement by introduced ones (Million; Asrat *et al*; Krüger *et al*, this volume). Thus, uncertainty about future policies generated by the experience of a highly interventionist approach to SWC has reduced people's confidence and willingness to invest in their own measures.

Tenure insecurity is another important factor which reduces people's willingness to invest in environmental management. Some argue that it is only with privatized land and exclusive tenure that people will be willing to make significant contributions to SWC. Some of the case studies provide support for this claim. For instance, in Malawi much greater investment in SWC is found on private farms and tea estates, compared with smallholder farms (Mangisoni and Phiri, this volume). However, the reasons for this do not lie only in the tenure system of the two areas. Neglect of the small-scale sector over many years has meant that people are required to work as short-term contract labourers on the tea estates and the large private farms, thus withdrawing labour from their own lands. In such dualistic economies, replicated elsewhere in such places as South Africa (Phillips-Howard and Oche, this volume) and Zimbabwe (Hagmann and Murwirwa, this volume), inequities in land distribution have undermined people's abilities to invest in their own land.

The introduction of new technology may result in changes in tenure regime as land changes value. For instance, in western Sudan, the introduction of earth bunds, particularly with land-moving machinery, has opened up opportunities for cash cropping by merchants and businessmen, resulting in the effective privatization of high-value *wadi* land (Yagoub, this volume). The flexible, traditional system regulated by community and tribal leaders was thus usurped, and local farmers, making use of the traditional *tera* system, lost out. By contrast, in the Red Sea Hills, in another part of the Sudan, access to land and water resources are tightly regulated by Beja tribal groupings, preventing outsiders and élites from taking over these assets (El Sammani and Dableub, this volume).

In some areas, SWC measures have been developed on common land which is managed and controlled by community groups. In such situations, effective enforcement of common property rights is vital for successful resource management. Key conditions for success include the existence of local organizations where a common set of purposes for resource management are agreed. Such organizations must be able to agree on a set of use rules, to exclude other users and to employ effective sanctions against 'free riders'. Relatively small cohesive groups with strong leadership appear the most effective, especially when they are managing resources which are perceived as valuable and easy to protect with relatively low transaction costs involved (Ostrom, 1990; Bromley and Cernea, 1989; Bromley, 1992). Several examples of common property management involving SWC are discussed in this volume. Individual and village-level initiatives in the Sahel have successfully rehabilitated thousands of hectares of barren land (Hassan, this volume). This process of investment in SWC has been accelerated by the increased clarity and affirmation of villagers' rights to manage their land and to control access to resources (Ouedraogo *et al*, this volume).

In general, there is no clear evidence of investment in land improvement being higher in areas where land title exists compared with where land is held under customary rights of use. It appears that the *de jure* system is less important than the *de facto* tenure regime, so long as secure land rights are guaranteed through customary tenure (Place and Hazell, 1993).

Access to information and technology

Options for SWC evolve with changing access to information and technology. A major development in many of the study sites has been the introduction of the plough and the increasing reliance on oxen for ploughing on the flat. This has reduced the level of investment in ridging or mounding techniques in many areas, but has provided new challenges for SWC (Lema, this volume). In other areas, however, the number of draught animals available per household is declining owing to changes in the environment through drought or disease, rising demographic pressure or external intervention (Mbegu this volume; Ostberg, 1986). In other cases, such as in the Matengo highlands in Tanzania, ox ploughing is not feasible because of the terrain and traditional methods for land cultivation have continued (Temu and Bisanda, this volume).

Elsewhere, the new forms of technology that emerge are combinations of previous practice and introduced innovation. For instance, in Zimbabwe farmers are experimenting with locally evolved pitting systems, together with tied ridges and adapted *fanya-juu* systems introduced by extension agents

(Hagmann and Murwira, this volume). Similarly, in Tanzania farmers are searching for mounding techniques which will reduce labour requirements and are compatible with plough agriculture (Mbegu, this volume). In eastern Morocco, as opportunities for high-value fruit and vegetable production open up on the plains through the harnessing of water, SWC techniques developed in the mountains are being transferred and adapted by migrants moving to the lowlands (Chaker, Chapters 4 and 5 this volume). The flexible combination of the old with the new draws on outside sources as well as generations of local knowledge and practice, and offers an important route to success. However, some traditional practices may not be relevant in today's settings; there is no point in deifying 'indigenous' techniques if they are no longer appropriate. Just as farmers reject externally imposed interventions that are inappropriate, so too will they abandon indigenous techniques. More important than the technique or technology itself is the process by which it arises; how different information sources and technological choices are derived from a little scientific experimentation and a lot of farmer practices. Building on tradition, rather than either replacing it or reimposing it, is the key to success. This is the subject of the next section.

Building on tradition: supporting indigenous SWC

Farmers across Africa have always understood changes in their local environment and assessed the problems they face. They have needed to design, select and adapt technologies in order to survive and prosper. Their ability to do so successfully is moderated by social networks and local institutions. External intervention adds another dimension to this process. In the final section of this chapter the question is asked: how can external interventions build on tradition and facilitate technological innovation by farmers?

Indigenous and introduced: complementarity or conflict?

As already noted, the distinction between indigenous and introduced technologies is, in many cases, artificial. Many so-called indigenous technologies have been introduced and virtually all introduced technologies have been indigenized through local adaptation. Indeed, it is this fluidity in the processes of technological change that is striking. However, when introduced technologies are imposed and prospects for local adaptation are constrained, problems arise. Many examples of this have already been mentioned and many more are discussed in the following chapters. For the development planner and project administrator, the appeal of an 'off-the-shelf' technical package is high. Simple diagnosis of a problem over wide areas means that a standard solution can be applied. Administrative procedures for delivery are predictable, equipment needs, labour demands and costs can be calculated easily, and schedules for implementation set, with predetermined physical targets and monitoring and evaluation procedures. The bureaucratic culture of government departments, development projects and donor agencies thrives on this type of procedural formality. Incentives for individuals and organizations are geared towards such outputs: targets for disbursement of funds, for food aid delivery and for project implementation must all be reached. Driven by the logic of internal planning frameworks, such projects have been, and continue to be, implemented on a large scale.

One of the big driving forces behind this style of SWC project in Africa has been food aid. Mechanical SWC appeared the ideal vehicle for the range of food and cash-for-work schemes that are popular components of food-aid distribution and employment-based safety-net programmes in areas suffering chronic food shortages. It seemed that all you needed was to peg-out simple contours, give people picks and shovels and everyone would benefit: food aid would be delivered, food shortages would decrease and long-term environmental investments would be created. Experience over the past decades has proved, however, that this is by no means always the case.

For instance, in Ethiopia massive investment in contour bunding through government and World Food Programme initiatives resulted in impressive structures being built. However, these were deeply resented by local people, and when the Derg regime fell in 1991, many were destroyed in protest (Dessalegn, 1994). With less authoritarian control in the rural areas today, many peasants are removing or redesigning these structures, to take up less precious land, to ensure that they do not harbour rodents and other pests, and to encourage soil and water to flow in desired ways.

The large-scale campaign approach to SWC is largely incompatible with

Characteristics of externally and locally derived SWC technologies

Characteristics	External	Local
Designed by	Engineers and development planners	Local farmers
Designed for	Soil conservation	Multiple, depending on setting (including soil/water harvesting, conservation, disposal)
Design features	Standardized in relation to slope features	Flexible, adapted to local micro-variation
Construction	One-time	Incrementally (fitting with household labour supply)
Labour demands	High	Variable, generally low
Returns	Long-term environmental investment	Immediate returns
Project setting	Large-scale, campaign approach; food-for-work/cash-for-work/employment-based safety-net programmes, etc	Longer term support to indigenous innovation; participatory research and farmer-to-farmer sharing

locally generated technology, as demonstrated in the table on the opposite page. Major differences exist in the objectives, design features, construction patterns and labour requirements. Externally derived technology is not necessarily inappropriate. Indeed, many indigenous systems are based on principles of bunding, contour strips and other introduced methods. More important is the manner by which the technology is introduced. Large-scale food-for-work programmes aim to mobilize large amounts of labour for simple tasks within a short space of time, an approach which is fundamentally incompatible with supporting indigenous SWC practice.

Participatory processes: building on local practices and supporting farmer-to-farmer spread

A participatory approach to rural development is, at one level, much more modest than the grand-scale campaign approach typified by the food-for-work schemes described above. But at another level it is much more ambitious, based as it is on fundamental 'reversals in normal professionalism'. Robert Chambers (1994, p xv) observes:

> Reversals imply a new professionalism. This is not a rejection of modern scientific knowledge, of research stations and laboratories, of scientific method. These remain potent, have their own validity and will always have their place. Rather, it is a broadening, balancing and up-ending, to give a new primacy to the realities and analyses of poor people themselves. These themes and insights are liberating for agricultural scientists and extensionists, opening up new ranges of experience and ways of working.

A number of the chapters in this volume describe participatory approaches in the development of SWC approaches. These approaches provide a bridge between indigenous and external expertise, with the resulting interventions often being interesting hybrids, drawing inspiration from multiple sources. Box 1.4 describes the participatory process followed in the project described by Hagmann and Murwira (this volume). Here, farmers have adopted low-cost irrigation technology for their gardens, adapted existing contour ridges using an innovation tested by a local farmer, and experimented with and subsequently adapted ridging technologies designed by research stations.

Examples of this type of process are expanding across Africa, as the limitations of the conventional technology transfer approach are increasingly being realized. The adoption of such a process is occurring in a wide variety of settings. This volume contains a number of different examples. For instance, in Zambia participatory research in the government agricultural research system is described (Sikana and Mwambazi, this volume); in Burkina Faso and Mali (Ouedraogo and Kaboré; Kassogué et al; Wedum et al, this volume) alliances between farmer groups, non-governmental organizations (NGOs) and the government are described.

The conditions for success are multiple, combining a conducive policy

---

BOX 1.4: A PARTICIPATORY RESEARCH, DEVELOPMENT AND EXTENSION PROCESS: AN EXAMPLE FROM ZIMBABWE (see Hagmann and Murwira, this volume)

1. Raising awareness, building confidence, developing group solidarity, using Training for Transformation or Development Education (DELTA) techniques based on Paulo Freire's approach to adult education.
2. Joint diagnosis of problems, exploring different views and perspectives among different groups of people (men, women, children, rich, poor, old, young), using a range of participatory rural appraisal techniques.
3. Exploring options through visits to other farmers and visits to research stations.
4. Testing out different ideas, first on a small scale. Farmer monitoring and review of the results through feedback workshops, cross-visits, etc.
5. Farmer-to-farmer sharing through informal networks and contacts, as well as through exchange workshops and visits.

---

environment, effective institutional setting, access to a range of participatory methods and approaches, and personal changes among researchers and development workers (Pretty and Chambers, 1994). The researcher and development worker must acquire new skills, new attitudes and new behaviours (Chambers *et al*, 1989; Chambers, 1993). Rather than planning, directing and enforcing, he or she must facilitate, convene, catalyse and negotiate. Rather than technological outputs, the focus is on the process by which technologies arise, become adapted and spread. Rather than dividing responsibilities between researcher, extensionist and farmer, roles combine and joint activities are central. These are big changes to the conventional, linear model of technology development. But they are proving successful, as many of the case studies illustrate. With the shift made from a high level of external intervention in design, planning and intervention, to a more facilitatory role, costs also drop, especially after an initial emphasis on training and local capacity building (Shah, 1994).

Successful participation therefore involves major reversals, certainly in professional behaviour and attitudes, but more fundamentally in power relations between different actors in the development process (Scoones and Thompson, 1994). This provides some very basic challenges for development organizations which wish to develop a participatory approach to SWC and natural resource management more generally.

CONCLUSION

The challenge of supporting more effective environmental management in Africa is huge. Three issues stand out from our earlier discussion of indigenous SWC and its ability to sustain the soil for the future.

First, we must be wary of simple definitions for a complex 'problem'. Much effort is expended on designing and disseminating 'solutions', but too little time is spent on understanding the problem. We have to ask from where our definition of a problem arises. What data are being used to describe the problem? What evidence is emphasized and what issues are ignored? What are the political interests involved in describing a situation? With more reflection on assumptions made by the various actors in the development process, more effective solutions will, hopefully, emerge.

Secondly, we must recognize that technology exists not simply as an engineering design, but in a social and economic context. For a technology to be attuned to people's needs, local environmental conditions and economic factors, it must be flexible and adaptable. Rigid prescriptions and designs do not work.

Finally, participatory approaches, while clearly desirable, are not uncomplicated. Simplistic adherence to 'community management' may mask important differences, between men and women, between young and old, between rich and poor. Any intervention inevitably affects the balance of interests within and between groups, with some winning and others losing. It is vital to appreciate the political consequences of development activity. This means recognizing conflict rather than ignoring it, so encouraging a process of negotiation and choice, involving all actors, in development planning and implementation.

2

MAKING THE MOST OF
LOCAL KNOWLEDGE

Water harvesting in the Red Sea Hills of Northern Sudan

Mohamed Osman El Sammani and
Sayed Mohamed Ahmed Dabloub

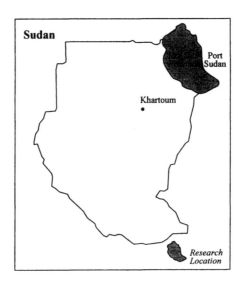

Indigenous soil and water conservation methods have been practised for centuries in the Red Sea Hills of Sudan. Hand-dug wells and earth embankments
provide a domestic and agricultural water supply which has enabled farmers
and pastoralists to exploit the limited potential for farming in the region. This
chapter looks at the continuing importance of indigenous techniques of water
harvesting in a region which would otherwise be unable to support human
habitation.

INTRODUCTION TO THE AREA AND PEOPLE

The Red Sea Hills occupy the north-eastern corner of Sudan, bordered to the east by the Red Sea. Covering a total of 219.920km^2, the region is dominated by mountainous, hilly or undulating terrain, interspersed with comparatively narrow valleys which are known locally as *wadis* (or *khors*). The western edge of the Great Rift Valley forms a high scarp made of old granite which runs from north to south about 20–40km from the coast. The scarp falls away towards the coast, opening up into some large delta areas.

Natural vegetation cover is limited to the low-lying lands on the coastal strip and *wadi* beds. The mountains and hillslopes are bare owing to the low rainfall received, the quick run-off and the lack of soil. Thus, the only plants which survive in this arid environment are desert and semi-desert types. The dominant tree species in the hinterland away from the coastal plain are samer (*Acacia tortilis*) and sayal (*Acacia raddiana*). Tree cover is generally denser at the edges of *wadis* and the well-drained depressions of good soils. They are a vital resource for those living in this inhospitable environment, as they provide shade, fodder for livestock, domestic fuel needs, as well as protecting the *wadi* systems from erosion by flash-floods.

The rural population of the Red Sea Hills is made up of four ethnic groups: the Beja, Beni Amer, Rashaida and the Fellata, as well as some refugees who have moved into the area more recently, mainly from Eritrea. The Beja are the dominant social group in the region, both in terms of numbers and the length of time they have lived in the region. Beyond the Gash and Tokar deltas in the low-lying coastal strip where irrigated agriculture is practised, the Beja depend on pastoral transhumance.

Transhumant pastoralism

The importance of livestock rearing in this arid zone stems from the role that animals play as the main providers of food, money and social status in the community. They are also best adapted to exploit the region. Cattle cannot survive the arid conditions of the Red Sea Hills region and are limited to the richer irrigation areas of the Gash and Tokar deltas where the vegetation cover is much denser. Goats, on the other hand, thrive on the limited vegetation in the area and are the most important producers of milk and butter, the two main items which, together with *dura* (sorghum), make up the population's diet. Most of the Beja families keep up to ten goats, and some families also keep sheep and camels.

Pastoralism is practised as a transhumance activity with households operating from fixed settlements along *wadi* courses. Rangelands are communally owned by the tribal groupings, each having rights over a specific locality. Normally, each tribal group owns a territory comprising a *wadi* system, with the wells dug for supplying water. Within the tribal land each household has equal rights to the rangelands. Family members, usually men, move with their livestock away from the permanent settlements to seasonal settlements where there is grazing. Grazing within the *wadi* environs by outsiders is permitted at certain times of the year, although these affiliated groups have no right to dig wells in the area. The Beja also recognize the importance of trees to the overall balance of the hydrology of the region: dead wood collection is allowed, but the cutting of standing trees is prohibited. Access to communal grazing land is thus strictly regulated by each tribal grouping and maintained individually by family members who make up the group.

Hand-dug wells

Permanent settlements along *wadi* courses are always characterized by the presence of hand-dug wells close to the settlement, many of which have been used since time immemorial by the Beja for water use in the home:

> There are five permanent wells at the Khor bed which contained a good supply of water. One of the wells was 53ft deep and contained 18ft of water. I was surprised to think that the local inhabitants had the perseverance to dig such wells, but the guide and the police volunteered the information that they were dug by the ancients with stone hammers. (Colchester, 1929)

Hand-dug wells are sparsely located, following localized aquifers, *wadis* and some deltas which receive surface run-off. Experience guides expert tribesmen to suitable sites for digging, mostly around grazing and settlement areas. They are then dug by members of the community using local tools to depths of 10–50ft, and are usually around 5ft in diameter. The lining is made according to the locally available resources, such as stones, bricks, crossed wood or tree branches and shrubs.

The durability of the wells is dependent on the nature of the aquifers, some tapping permanent and rich basins, while others are found on temporal, sub-surface water formations. Salinity is a major problem, in particular at sites closer to the sea where the water extracted is fit only for livestock. Annual replenishment by surface run-off is essential for reducing salinity levels in the wells and for generating adequate supplies. Without the wells, permanent settlement would be impossible.

Beja agriculture

The Beja have integrated farming into their pastoral way of life, growing sorghum and millet on small plots of flat land along *wadi* beds. Thus, the distribution of population along *wadi* courses is not only determined by access to grazing resources and the presence of wells for domestic water, but also according to the availability of agricultural land, the existence of marketing centres and accessibility to transport routes. Silty clay loams are found mostly on the edges of *wadis* and are currently the preferred soil for crop growth owing to their high water-retention rate, besides being easily worked by the farmer.

The most important factor in determining an agricultural site is the geomorphology of the *wadi* which determines the availability of suitable land for cultivation and the occurrence and intensity of flash-floods.

Flooding depends on two seasons of rainfall, a winter period from October to February and a summer period from July to October when rains fall. The winter rains are vital for plant growth, while the summer rains extend the growing season for plants. The Beja plant a long maturing variety of sorghum which normally needs four months to grow, and the sowing depends on the intensity of the flooding – the more land under flood, the greater the extent of sorghum under cultivation.

While the two seasons are virtually guaranteed, the amount of rainfall is highly variable and patchy, particularly in the north and east where many areas may go for years without rain. The region suffers regularly from widespread drought, such as that experienced in 1984 in which the floods failed in most of the major *wadi* systems and rainfall was well below average. It is difficult to assess the impact of such enormous variations in rainfall on farming, but it can generally be stated that out of every five years, there are on average only two seasons of cultivation.

Harnessing the water

The most obvious way to control water running through the extensive system of *wadis* is to dam the flow, redirecting the water to irrigation schemes on either side of the *wadi* bed. The problem with this, as was discovered in the late 1950s when a diversion dam was built at El Garad, is that *wadis* regularly change their course. Fifteen years after the dam was built, the *wadi* moved, leaving behind a static dam which is now quite a considerable distance from the current *wadi*. Dams also cannot cope with the sudden flash-floods which accompany the seasonal rains. Instead, the Beja have developed a more flexible range of soil and water structures which are much more effective than dams at capturing the surface run-off from the *wadis*.

SIMPLE LOCAL TECHNOLOGIES WHICH WORK

The techniques applied are simple, based on local technologies which have been handed down through generations and using whatever materials are available. The most common method is the earth embankment built across *wadi* beds. This succeeds in spreading the water outwards from the gully and on to the fields beyond.

Each embankment is different, constructed to fit in with the shape of the gully and the intensity of water running through it. Large crescent-shaped embankments are built on *wadi* beds which are not too steep or deep and where the water runs in different channels (see the figure on next page). The water is

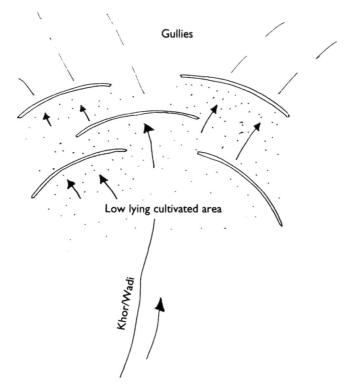

Low-lying crescent embankments

captured within each crescent which is then planted with sorghum. These embankments are low-lying, only about 100cm high, but encompassing an area of some 200 feddans (2.4 feddan approximately 1 hectare). As well as a variety of cross embankments, lateral embankments are built which run along the edges of the *wadis* to concentrate the flash-floods on to cultivated land. The cross structures are often found in combination with the lateral embankments to maximize surface run-off from the *wadis*.

In addition to these major structural works, the Beja farmers construct a variety of smaller structures on their plots. Through levelling the land and working the soil, a higher level of water infiltration is achieved. The soil thus benefits from the construction of water conservation structures through the deposition of silt and the working of the surface by levelling and the breaking of clods, improving the soil texture and by the repeated application of organic matter.

Labour requirements

There is no overall conflict between labour requirements for farming and live-stock management as herds are small in size and require little supervision. Farming is also carried out primarily during the rainy season, a time which requires minimal herd management. The system thus has a close 'fit' with the Beja social structure.

The labour requirements for the construction and maintenance of these soil- and water-conservation measures vary considerably from site to site owing to differences in the size of the land to be developed and the volume of the work needed. Major works are usually executed communally under the guidance of the sheikh. On privately owned land, members of the extended family work under the supervision of its head. In each case, earth is dug at the site using spades, shovels, axes and baskets, and vegetation is planted on the embank-ments and bunds to stabilize the structures. Stones and, on occasion, plastic sacks may be filled with sand to add to the stability of the embankments. Most of the maintenance of these structures is carried out before the rainy season begins, but during the rainy season flash-floods frequently breach the embankments and wash out terraces, which then need repairing.

Social organization

The farming communities are small and are based on tribal groups with land-holdings of less than 3 feddans and, unless they are accompanied by a hand-dug well which allows permanent settlement, they are often only occupied season-ally.

The essence of all the soil- and water-conservation practices carried out by the Beja is that it is a community activity, whether practised by a sub-tribal group or an extended family. Even those who do not own land and are associated in one way or the other with the group, secure the right of cultivation through paying a token to the owner of the land. Hence, differentiation according to wealth or right of ownership to farming land plays no part in the acquisition of land in the Red Sea Hills region. The solidarity of the group and the established right to the land permit all adult male members of the household to practise farming when they choose, while absentee heads of household entrust their relatives to culti-vate on their behalf. Women have specific roles in family life, such as processing milk, butter, ghee, etc, fetching water and firewood, and caring for small and sick animals in the home. In general, they have a low status compared to men, who completely dominate them, and they are not involved in farming or building the soil and water conservation structures.

Initiatives to develop water

Wadis clearly present the most viable land and water resources for development in the Red Sea Hills area. Not surprisingly, there has been a continuous government interest in utilizing them for the improvement of community life in this area, with some programmes bringing greater success than others. The objectives of these schemes were many: to improve the water supply, to increase sorghum production and to create centres where services could be provided.

One of the most prominent government schemes operating in this region is the Soil Conservation, Land Use and Water Programming Administration (SCLUWPA). This began as a government programme in 1986 for the promotion of soil and water conservation techniques over the whole of the Red Sea Hills area, starting in places where such practices are well established and expanding into new areas in order to increase sorghum production for household consumption and cash sales. Its approach is to build on people's acquired knowledge of soil and water conservation, and its success is largely dependent on community participation in such a scheme.

Since SCLUWPA began its work in 1986, there have been increases in the acreage reclaimed for agricultural purposes, the number of farming families covered, the provision of machinery tools and input supplies, and the drawing of women into the farming and development programmes. A 'Women in Subsistence Agriculture' programme aims to empower women and upgrade their capabilities in land-reclamation techniques, water management and agricultural practices.

On the technical side, improved designs, coupled with the use of advanced machinery and the provision of hand-tools have increased the capacities of structures and facilitated community participation in the reclamation and maintenance of soil-conservation structures. Dams and earth embankments have increased the efficiency of current soil and water conservation practices.

CONCLUSION

Indigenous soil and water conservation activities are gaining momentum in the Red Sea Hills area. As this chapter clearly demonstrates, indigenous methods of water conservation are crucial to farming in this arid environment, providing the most feasible way of irrigating crops using simple technology and locally available materials. The farming system evolved to date is integrated in many aspects into the Beja way of life, both in ecological and economic terms. The future is likely to witness the expansion of cultivated areas, improvements to the techniques applied and the increasing involvement of farming communities with new soil and water conservation techniques.

3

DROUGHT AND THE NEED TO CHANGE

The expansion of water harvesting in Central Darfur, Sudan

Yagoub Abdalla Mohamed

Central Darfur is one of the most drought-affected regions of the Sudan. The drought years of 1983–85 greatly affected the demographic and socio-economic conditions of the area. Large numbers of people left their homes owing to famine and the environmental effects of desertification and drought. This was accompanied by tribal conflicts, the growth of shanty towns and changes in the pattern of livestock raising and agricultural production. During this time, most people lost over half of their cattle, as well as large numbers of sheep, goats and camels.

This chapter looks at the province of El Fasher, a region which received a large influx of people from other drought-stricken areas of the Sudan. In order to cope with the incomers, rain-fed agriculture decreased and the indigenous technique of *wadi* cultivation through water harvesting has spread to all areas.

INTRODUCTION TO THE AREA AND PEOPLE

Central Darfur is situated on the northern transitional margin of the Inter Tropical Convergence Zone. Consequently, most of the area is deficient in water even in the wettest months of July to September, when over 80 per cent of the rain falls. During June, the hottest month, temperatures regularly reach over 45°C and in January, the coldest month, temperatures reach 18°C.

Prior to the drought of 1983, El Fasher had a population of around 283,800 (1983 census). In 1993 the population had jumped to 582,090 out of which about 362,100 were sedentary and the remainder were migrants who moved with their livestock looking for suitable grazing and farming areas.

This rapid increase in population resulted in severe pressure on natural resources in the region. The original tribal groups of the region traditionally combine crop farming with livestock raising, while the recent migrants from the Zaghawa tribe combine pastoralism with limited farming.

The distribution, amount and intensity of rainfall available for agriculture depends on the topography of the area. The eastern slopes of the Jebel Murra, well drained by *wadis* and gullies, form the catchment for three tributaries (Wadi Kutum, Wadi El Kej and Wadi Abu Hamra) that eventually join Wadi El Ku, the main river in the area.

The flood plains of the three main tributaries and the banks of Wadi El Ku are rich in alluvial deposits, from fine silts to sandy clay loams. The frequent deposition of sediment makes *wadis* change their channels, leaving behind flood plains which are suitable for agriculture. The bed of the flood plain is farmed when the *wadi* water recedes, while most of the traditional water harvesting techniques and flush irrigation are found on the fertile loamy soils which have good water-retention qualities.

Clay plains, mainly found in the areas between *wadis*, are thought to be the result of sheet-floods. The plains occupy large areas, although in many places they are buried by sands or eroded by gullies that form in the area. The slightly cracking clays are characterized by low permeability and high salinity and are not used very much for agriculture, except during good rainy years.

Sand-dunes cover the eastern side of Wadi El Ku. These deposits are consolidated and stabilized by vegetation, in particular the gum Arabic tree (*Acacia senegal*). These sands are relatively porous with coarse grains that absorb any rain that falls and store it until it is exhausted by growing plants or evaporation. The sands are used extensively for millet and sesame cultivation. However, in areas where grazing or cultivation has restricted tree growth, there are growing signs of desertification and sand creep is apparent.

Shifting cultivation

In order to minimize the risk of farming, most farmers practised shifting cultivation on both sandy soils and clay alluvial deposits. The sands were used to grow millet, sesame and groundnut. This was where the main family farm would be located, providing the bulk of family subsistence needs, while the clay soils were cultivated using water harvesting techniques to grow sorghum, vegetables and the cash crop tobacco in some cases.

TRADITIONAL CONSERVATION METHODS

The traditional water conservation method of Central Darfur involves harvesting run-off water by constructing low earth bunds called *trus* (*tera*, singular). The technique probably has its origins in the home garden or *jubraka*, a backyard farm operated by women growing quick maturing crops and some vegetables like okra, pumpkins and cucumbers. When villages grew in size and more livestock were kept at home, women began to extend *jubraka* activity to areas away from home. Hence, they started to grow the same crops under *trus* in small areas, while the large family farm in the sandy soil used to provide the subsistence needs of the family.

In recent years, as rainfed farming on the sandy soils has become increasingly risky and unable to produce enough food for the family, staple crop production has had to take another dimension, using other sources of water to increase soil moisture. The indigenous *trus* cultivation provided the answer.

How the *trus* system works

The bund traps run off generated after rain storms on catchments usually two or three times the size of the cultivated land (see the figure on next page). In 1964 the first local attempt to construct a large *trus* embankment across the Wadi El Ku succeeded: the farmers harvested a good crop of sorghum. This proved to the villagers living in the increasingly arid area that the only water source that could be used with some degree of certainty was harvesting surface run-off. For these reasons, many farmers began to shift to clay soils to practise *trus* cultivation.

Constructing *trus*

The technical design of *trus* is simple. A U-shaped earth bund is constructed with the top facing the slope and the catchment area. Any surface run off is thus trapped by the *tera*. The most complicated part of building a *tera* is to achieve the right 'fit' between the slope, catchment, direction and size of the *tera*. This is not achieved easily and farmers spend a great deal of time fine-tuning designs. A typical *tera* has a bottom earth bund of a certain length, determined according to available land and labour with upsloping arms. Because of the intensity of the

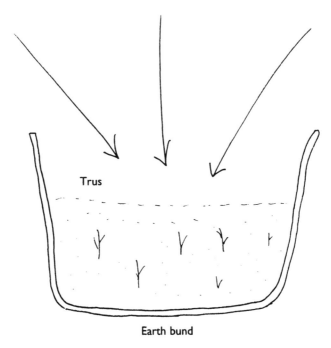

Trus

Earth bund

How the *trus* system works

labour required to build such embankments, the cultivable area tends to be small – between 2 and 3 acres in size.

Water spreading

In recent years, urban farmers and merchants have begun to be involved in what are termed water-spreading projects (see the figure on opposite page). *Wadis* are diverted using an earth embankment and small canals to spread the water behind the dam and on to agricultural land. The two systems (traditional *tera* and modern water-spreading projects) are both found in the area, but in different locations and tapping different water sources.

The main differences between them are summarized in the table on page 40. Thus, while the traditional *trus* cultivation originally evolved without outside intervention and proved its sustainability, the majority of modern water-spreading projects depend on the use of earth-moving machinery and high capital inputs. Private sector projects are proving to be very successful and show encouraging results, while government or co-operative projects tend to suffer from managerial problems and lack of access to machinery.

The large numbers of actors involved in traditional *trus* cultivation and modern water harvesting techniques demonstrate their success as locally

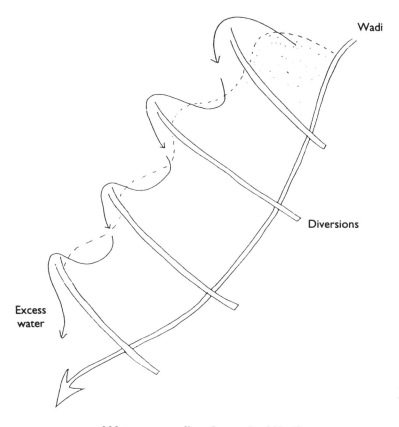

Water spreading from the Wadi

developed water harvesting techniques which have recently been revived (see the table on page 41).

The advantages of *trus* cultivation

The large number of people who have taken up *trus* cultivation reflects the potential of this water harvesting technique to solve the problems of food insecurity in Central Darfur. This technique is less susceptible to low and poorly distributed rainfall. *Trus* cultivation also results in higher yields per acre than rainfed agriculture, with an average acre producing five to seven sacks of sorghum, far more than the two to three sacks of millet obtained in sandy soils in normal years. A major constraint on farming in the region is labour availability. On the sands, weeding must be done very soon after the rains to enable any crops to make the best use of the soil moisture. The alluvial clay soils are much heavier to work than the sandy soils and hence demand more labour, but such soils also retain moisture and retain their fertility much longer.

Comparison between traditional and modern water harvesting techniques

Item	Traditional water harvesting	Modern water spreading
Source of water	Rainfall: sheet flow	Rainfall and *wadis*
Method	Construction of earth bunds and planting behind	Large earth embankments and planting in wet soil behind
Construction	Hand-tools and labour	Earth-moving machinery
Crops grown	Sorghum and vegetables	Sorghum, tobacco and vegetables
Purpose	Mainly for subsistence, vegetables for cash	Mainly commercial
Ownership	Small farmers	Merchants, urban farmers and co-operative societies
Farm size	2–5 acres	50–200 acres
Distribution	Mainly in gently sloping areas	Mainly along *wadis*
Cropping pattern	Only during the rainy season	Two cycles, rainy and winter
Potential problems	Rain failure, growth of weeds	Lack of machinery

Water spreading and *trus* cultivation have also given farmers the chance to diversify crops, giving them better opportunities to increase their income and to improve family nutrition. Water harvesting has extended the growing season to include the winter months up to March. This enables farmers to plant crops such as tomatoes, water melons and cucumbers which are harvested in June or July (off-season), allowing farmers the chance to benefit from the higher profits at this time. Lastly, the trend towards water harvesting on clay soils, if it continues, will give sandy soils the opportunity to recover from the effects of desertification.

The disadvantages of *trus* cultivation

Despite the successes achieved, there are a number of disadvantages associated with the broad take-up of water-spreading techniques.

Loss of common property
Trus cultivation evolved under customary land tenure with secure user rights. Local inhabitants have their traditional and customary rights to land that regulate use. Tribal leaders and village heads (sheikhs) are responsible for land allocated and regulating its use.

However, with the expansion of this and water spreading techniques in the

Actors involved in *trus* cultivation

Small subsistence farmers	The largest group involved in *trus* cultivation. Small farms, growing food crops and using heavy labour inputs from the family or work parties to ensure food security. Traditional *trus* cultivators use simple tools for bunding and weeding. Most grow millet under rainfed conditions on the sandy soils and sorghum in clay soils using the *trus* system.
Women involved in subsistence cultivation	Women have full control over their vegetable farms which are grown using the *trus* system. The vegetables may be used for family consumption or sold in the market. Such farms face some competition from vegetables grown in large water-spreading projects.
Small commercial *trus* cultivators	In some villages the *trus* cultivators have specialized in the cash crop tobacco, using the cash to buy food crops.
Merchants and urban farmers	Merchants grow sorghum and vegetables to sell, while urban farmers combine both subsistence crops to ensure family food security and the sale of vegetables.
Co-operative societies	These grow subsistence crops through water spreading techniques. Embankments are made by mechanical means. Through being members of a co-operative society, poor farmers are able to utilize *wadi* floods rather than depend on sheet flow.
Government involvement	These are either under the direct control of the government or distributed in the form of tenancies to local farmers. Farmers are given incentives such as land allocations or the distribution of improved seeds and some extension services.

area, major changes took place which affected the basis of the traditional land tenure system. Four main forms of land ownership are currently being practised:

1. Ghifar: community or tribal ownership of unoccupied land. This is used for grazing, gum Arabic collection and as reserve land for future farming activities.
2. Hakura: family- or clan-owned land. This emerged out of the tribal land because some families or clans were given large areas under their control by sultans. Such individuals control the right of use and the right to allocate the land to others.

3. Owner-like possession: a plot allocated by the sheikh to an individual, giving them all rights over it, including renting, selling and inheriting the land.
4. Government lease-hold land: a result of government intervention in allocating areas to investors coming from outside the local community, mainly from the towns.

The present diverse variety of *trus* methods of cultivation was able to evolve as a result of the flexibility of the indigenous land tenure system and its ability to accommodate social, economic and environmental changes. By contrast, the construction of earth bunds and embankments has created a trend towards land privatization. Building earth bunds and other physical structures leads to the permanent ownership of land, a fact favoured by recent migrants and urban farmers who would not have strong rights to land under customary rules of tenure.

However, tribal leaders favour customary tenure because it ensures that the rights of the community are maintained rather than losing all power to individuals. This conflict of interests has a negative impact on the development of water harvesting techniques. Migrants in particular have been denied access to *trus* land, mainly because the original inhabitants of the area fear that those migrants may claim ownership of the land which they develop and cultivate.

These trends have created conflicts between traditional organizations responsible for guarding common property and individuals seeking to establish private claims. Common property may increasingly be lost and the poor may become landless if this trend continues.

More involvement in growing cash crops

Trus cultivation brought with it more involvement in the production of vegetables for sale in urban markets. Despite the increased income generated, the trend, if expanded, may lead to shortages of staple food supplies, particularly given the high risk associated with staple food production on the dry upland sandy soils.

Reduction of downstream flows

The *wadis* in Central Darfur are the lifeline for communities living on their banks. They replenish wells, hafirs and dams. Large-scale water spreading projects affect the flow of these *wadis* and reduce the amounts of water downstream. Hence, communities living furthest from the source of the *wadis* may be badly affected by the development of *trus* and water spreading techniques further upstream which divert water for agricultural purposes.

CONCLUSION

The case of *trus* cultivation in Central Darfur demonstrates both the potential for the success of indigenous water harvesting techniques and also some of the pitfalls that such success can bring. The increasingly commercial aspect of *trus* cultivation potentially clashes with the local cultural context, in particular customary land tenure. If it is to continue to expand and solve local problems of food security, these issues will need to be addressed.

Nevertheless, *trus* cultivation is part of the local culture which has survived because it is technically known to all. It has also survived because it allows farmers to grow their staple diet, millet, in the sandy soils, and sorghum as a secure crop in the clay soils. The complementarity of clay and sandy soil cultivation maximizes the potential of this ecological setting and provides more secure food production in the region. Indigenous water harvesting is a positive element within this setting but one which requires careful support to maintain a balance between common interests and private gain.

4

THE MASTERY OF WATER

SWC practices in the Atlas Mountains of Morocco

M Ait Hamza

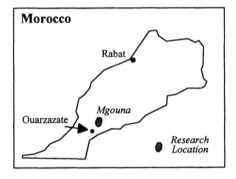

INTRODUCTION TO THE AREA

In the Assif Amgoun Basin in the Atlas Mountains of Central Morocco, arable land is scarce. Although overall population density is low, much of the terrain cannot be farmed and each hectare of cultivable land must support an average of 28 people. Because of its exposure to the Sahara and high altitude, the zone is characterized by scarce and irregular rainfall and a wide range of daily and seasonal temperatures.

Natural vegetation is limited to a few junipers, ilex and herbaceous plants. The extreme dryness of the air, the winter frost, the dramatic fluctuation in daily temperatures and the violence of the summer storms all subject the area to intensive soil erosion, as can be seen in the numerous accumulations of fallen, jagged rocks on the slopes, or the rocky debris and pebbles which litter the *wadi* beds.

Patterns of land use

Areas favourable to agriculture are scattered and patchy, coinciding with the presence of surface water in the *wadis*. The irregularity of water supply has led to a complex system of water management and distribution between farmers. Water and soil are the two determinants of human occupation within the landscape and generate a stark contrast between the lush ribbon formed by the oases and the vast empty spaces in between.

Three forms of land-holding are found within the area: the system of private land ownership (*melk*), which is by far the most common; a few small *habous* affiliated to mosques which exist in every settlement; and collective ownership which is much rarer. Cultivable agricultural land represents barely 0.8 per cent of the total surface area of the basin, and the average land-holding area is no more than 0.7ha.

The very small size of holdings is a reflection of the physical constraints and the extremely limited amount of land available in relation to demographic pressure. The different plots of a property are all extremely narrow and far apart. A holding of 3ha can be divided into 60 plots, spread out over three different settlement areas 30km apart. Cultivable land is maximized by patient maintenance of terraces.

In order to survive, farmers pursue a wide range of non-farming activities, such as gathering forest fruits and, more recently, work linked to sport and mountain tourism (eg guides, porters, hostel keepers, musicians). No cultivation is possible in this part of the Atlas without irrigation. While the sparseness of the soil imposes limits on the spread of cultivation, water is also a determining factor, and mastery of water lies at the heart of community life.

Diverse incomes: crops, livestock and migration

The basic crops are barley, wheat and maize, grown alone or in mixed stands. Trees do not receive any particular attention and must compete with other crops for water, fertilizer and care. The main species are fig, almond, apricot, peach and walnut, although pomegranate, vines and apple trees can also be found. Today, most of the big walnut trees have been cut down. Rose bushes are planted on the borders of fields or alongside irrigation channels, where they form thick hedges.

The livestock kept must depend on fodder from forest and scrubland. Grazing areas have been overexploited and have suffered from drought and erosion, making it increasingly necessary for farmers to buy fodder, such as barley and bran. Flocks consist of local species which are hardy but not very productive. Livestock and crops are closely integrated, each activity benefiting the other. Large flocks are rare, the average flock consisting of 50 small ruminants and 2.4 cattle. While it is unusual to find a flock of more than 400 head, it is also rare to find households without any animals at all, if only a few sheep and goats.

Since the 1950s, this area has become a major source of migrants. This tide of migration has created an almost total dependence on the towns and countries to which they have moved. The old socio-cultural, political and economic values have been turned upside down by the money injected into the region and the attitudes the migrants have brought back with them. Emigration has had a considerable influence on consumption, as well as on production techniques. It has also favoured a new dynamism in political and social arenas. The emigrants take part in elections and support community developments such as electrification, the building of mosques, the construction of roads and welfare facilities. Above all, migration has given farmers the means to rent or buy new equipment, tools and land.

TRADITIONAL METHODS OF SOIL AND WATER CONSERVATION

There are two traditional techniques for soil and water conservation: hillside and fluvial terraces (see photo 1). Hillside terraces are aimed at creating easily cultivable areas by breaking the slope, forming soil on the shelf so built, and conserving irrigation water by reducing its speed of run-off. The width of the terrace depends on the slope and depth of the soil, but it tends not to exceed 10m in order to minimize the distance the earth has to be transported and the height of the wall. In high altitude areas, such as *Ouzighimt*, dry-stone terrace walls have been interwoven with tree-trunks. The abundant use of juniper trunks in the building of terraces, some of which are over 100 years old, is evidence of how rich the forest cover once was. There are only a few stunted relics of these junipers now and dry-stone walling is used more and more. In the mid-mountain areas, there are larger or wider terraces where the slope is gentle and water plentiful. Terrace walls are built of dry stones, strengthened by the roots of trees (vines, almonds, walnuts) planted along the top.

Fluvial terraces help to protect fields and crops along the *wadi* from flooding. The steep and narrow river-bed generates a powerful flood, resulting in either the complete erosion of the terraces or the destruction of the fields by gravel deposits (see photo 3). Hedges are planted along the rivers and streams in order to prevent the flood waters from submerging the new terraces, and act as a filter for large deposits, only allowing through the finer particles which act as a fertilizer for the soil. At the same time, these hedges provide wood for cooking, building and furniture-making. The trees also create a micro-climate encouraging the growth of fodder. The type of trees grown have to be adaptable to silty sandy soils and require little maintenance. Willows, tamarisks, white and black poplars, reeds, laurels and brambles are the most popular species.

New techniques introduced

Since the early 1960s these techniques have undergone certain changes. Cement and gabions have been introduced by the authorities and the Ministry of

Agriculture, but these dykes do not withstand the violence of the floodwaters very well. They are too compact, have weak foundations and are not well constructed. The farmer is not involved in the design of these structures and is therefore not very interested in the efficiency or durability of the technique, while the large-scale introduction of wages to pay for the construction of the dykes has contributed to a decline in the traditional communal support system of mutual aid.

SWC techniques are indispensable in narrow valleys and for irrigation, but they also have the disadvantage of creating a landscape which is composed of long and narrow plots, which make it impossible to mechanize agricultural operations. The surface area taken up by the terrace walls and protective plant barriers is very large and accounts for a major loss of space in a region where arable land is at a premium. Nowadays, sons of poor families who have emigrated and acquired money have been able to buy land. Their interest in agriculture has as much to do with social affirmation as with crop production. The former rich, land-owning families are now suffering competition from migrants and the lack of cheap farm labour.

There appears to have been no significant abandonment of traditional techniques, and in some areas there has even been limited expansion. However, these very narrow mountain valleys have probably reached the limit of their development potential. The farmer expends all his energies in preserving the soil which has been reclaimed from the flooded *wadis* and streams.

CONCLUSION

Traditional SWC techniques, developed in the Moroccan central High Atlas, represent an adaptation to harsh mountain conditions. Their main functions are to stem erosion and to create irrigable arable land on the steep and inhospitable slopes. The construction of man-made terraces demonstrates the ingenuity of this people and explains their close attachment to these small plots which have been built up over generations. The investment of effort in these patches of land is out of all proportion to its economic yield, but its social value is inestimable.

5

MOUNTAINS, FOOTHILLS AND PLAINS

Investing in SWC in Morocco

Miloud Chaker with H El Abbassi and A Laouina

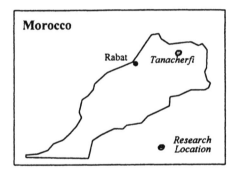

The area covered by this chapter lies between the Horst mountain range and its hinterland to the north, which is made up mainly of low plateaux and plains where the climate tends to be arid. The mountains, whose highest peak is barely more than 1700m, receive an annual average rainfall of 450mm, and the foothills get no more than 300mm. These two different environments complement one another. Semi-nomadic herders, for instance, have long been able to exploit grazing land in the highlands and foothills alternately, while the highlands have rich potential for irrigated agriculture as well as animal husbandry. The highlands also provided security for the population during times of conflict and political crisis. This study looks at how agriculture has been changing in mountain and lowland areas, the persistence of traditional methods to conserve soils and moisture, and the impact of new technology, such as irrigation pumps.

INTRODUCTION TO THE REGION AND ITS PEOPLE

The region is inhabited by Berber people who have long been settled there. The Chorfa, who are not Berbers, were the last to arrive in the region, and occupied the most hilly terrain as they could not find adequate land elsewhere. They have developed the most impressive SWC techniques to cope with such difficult conditions and are now transferring these techniques to the foothills where they have been able to purchase new land. After the harvest, farmers get on with improving their fields by removing stones and constructing embankments, while some may leave the village to look for seasonal work in towns or on irrigated farms.

According to the 1982 census, the Tanacherfi commune, which includes the area in question, had a population density ranging from 10 to 20 inhabitants per km^2. In the foothills, where land is relatively abundant, the average is 2.4ha of cultivable land per inhabitant – ie, 25ha per household. In the highlands, by contrast, one irrigated hectare has to feed 15 people, exerting a pressure on land which helps to explain the maintenance and extension of traditional SWC structures. In the foothills, SWC techniques are mainly designed to stem active and obvious erosion, while in the highlands, flat plots of land must be secured and maintained by terracing, permitting irrigation. SWC structures help to store rainwater in the soil, but must enable excess water accumulated behind the terrace to drain away, whilst fostering the growth of trees planted along the structures.

Socio-economic trends and rural development

In the highlands, irrigated agriculture has a very long history, these mountain areas providing diverse resources for varied crops to be grown and giving access to abundant water, pasture and firewood. However, the cultivated area was limited and forced people to start farming on steeper slopes. SWC techniques were designed to create areas for cultivation rather than just to conserve soil and water (see photo 2). Although the highland people were still semi-nomadic, irrigated farming and tree cropping demanded that part of the farm be settled while the herds could move with other members of the family to the foothills when times were hard. Such movement is now extremely limited as more people have settled permanently. However, the relationship between the highlands and lowlands remains very active and close, as almost every household has land in both locations. The household is often divided into two parts, one in the highlands and the other in the lowlands, each part gradually evolving into separate independent households. In this way, the traditional techniques long practised in the highlands have been transferred to the foothills by the people who have come to settle there permanently.

Twenty per cent of the population has migrated abroad or to other parts of the country, in most cases looking for work. However, attachment to the village remains strong as workers abroad often dream of retiring there. There has been a large increase in land prices, accompanied by the introduction of mechanization and new farming techniques. Land has become increasingly expensive, even though yields remain poor and uncertain from one year to the next. In the 1950s, 1ha cost between 1000 or 1500 dirhams (US$1 equals 8.5 dirhams); nowadays the price varies, according to the type of soil and the location of the plot, from 10,000 to 20,000 dirhams per ha.

Land tenure

All land on which traditional SWC techniques are practised is privately owned. On the other hand, grazing and scrubland in the mountains since 1930 have been the responsibility of the Forest Service, and the cultivated area may not be extended there. Farmland, therefore, is extremely limited in the highlands and increasingly fragmented as a result of the growing population. Some holdings are made up of more than 20 plots which are scattered throughout the highlands and lowlands. The plots closest to the house tend to be well maintained and conservation techniques best developed, whereas plots far away from the house are usually rather neglected.

The scattering of plots is one of the reasons for the introduction of agricultural machines; ploughing and harvesting can now be done in a few hours instead of weeks, and more than 50 per cent of land in the foothills is now ploughed by tractor. Mechanization demands that stones be removed from fields and gullies filled in, the stones being used to construct bunds and rock lines across steep, gullied slopes. Given the rapid spread of machinery, these techniques are likely to become more and more common. All farmers are now cleaning up their fields and setting up stone lines and embankments to retain water and eroded soil.

Farming systems

In the highlands, farmers specialize in producing fruit and vegetables. Cereal crops are secondary and are only grown along the edges of basins or where irrigation water is lacking. Given their relative isolation, agricultural production has been designed primarily to achieve self-sufficiency and only surplus production is sent to market. Irrigated plots are intensively farmed: on the edges, along the retaining walls, almonds, figs and vines are planted, while vegetables, such as potatoes, turnips, carrots and onions, are grown under the trees and in the middle of the plots.

Tree cropping is currently gaining ground owing to the competition which is now faced from the produce of irrigated farms in the surrounding plains. Looking after trees, especially almonds, requires little expenditure, so the produce remains competitive in comparison with the plains, whereas annual crops involve heavy costs. Since it is difficult to transport crops from the highlands to the market owing to poor communication and transport networks, costs can take up to 30 per cent of the selling price of vegetables, whereas shelled almonds command a high price with minimum transport costs. Some plots are now completely planted with trees and annual cropping has been abandoned.

Highland dwellers are also increasingly producing fodder crops to fatten livestock which remain the most secure investment for local people. Goats make up 80 per cent of livestock numbers as they can make use of common grazing land within state forest areas. But when times are hard in the foothills, animals are often driven to the mountains to make use of common grazing, thereby doubling the stocking density. The best maintained plots and conservation measures are found where irrigated cropping is associated with animal husbandry, since this enables plots to be well fertilized.

In the foothills, rainfed farming is the main activity, combined with extensive animal husbandry. Barley accounts for 80 per cent of the cultivated area, the remainder being given over to wheat and potatoes. Cereal cropping is primarily designed to meet local needs. Livestock feeding in times of drought relies heavily on the farmer's stock of barley, oats and straw and residue from the beetroot crop purchased from elsewhere. In this rather arid environment neither rainfed farming nor animal husbandry on its own can provide a secure income, and it is only by combining herds and crops that farmers can overcome drought and uncertain yields. The latter range from nothing during dry years to more than 1200kg per ha in years with adequate rains.

While farmers might have little incentive to conserve water and soil given the low grain yields, the fodder value of these cereals can be very high. Although the value of 100kg of barley is only 120 dirhams, the same quantity of barley can fatten more than five lambs, which will bring in much more. In the study area, households have on average 45 head of livestock. Sheep are renowned for their profitability (800–1000 dirhams for a lamb) and farmers' strategy is to produce as much fodder as possible and store it for the hard times ahead.

TRADITIONAL CONSERVATION MEASURES IN EASTERN MOROCCO

In the study area, the measures used are both biological and mechanical. The most solid, long-lasting and common structures are built of dry stone and then reinforced by planting a line of trees along the retaining wall. In the irrigated area, SWC structures are very sturdy and well maintained, whereas in the rainfed farming area, they are no more than discontinuous dry-stone embankments, usually far apart and low in height.

The height of the retaining terrace wall is determined by local conditions such as the slope, the availability of construction materials and the type of land use planned. In the highlands, stone walls may be over 2.5m in height and sometimes several hundred metres long. The height of the retaining walls is designed to level the plot so that it may be watered properly by irrigation canals (see photo 3). The length of the walls is determined by the topographical features of the plot to be improved and by the land tenure situation, as it is not possible to build continuous structures on fragmented plots. Most terrace walls have been built in phases as they must be raised continually to avoid overspill of the soil which is constantly carried down the slope and builds up just behind the wall. Only where there is no slope at all is it possible to stop raising the height of the structures. The work therefore represents the accumulated efforts of several generations, which makes it difficult to assess how much work has gone into establishing such structures.

BOX 5.1: TRADITIONAL SOIL AND WATER CONSERVATION IN THE EASTERN PART OF THE RIF MOUNTAINS

The eastern part of the Rif Mountains forms a zone of transition and is characterized by low rainfall (400mm). In the mountains and foothills of the Boudinar basin, the average population density is over 100 persons per km^2. This is very high, because 14 per cent of the total area consists of uncultivable, bare land. In the 1960s and 1970s considerable outmigration occurred, mainly of young men, while today entire families are emigrating.

Agriculture and livestock are the main economic activities in the Boudinar region, with 76 per cent of the total area under crops. This is only possible because farmers have constructed numerous forward sloping terraces, step terraces and stone walls to create fields on the steep hillsides. Check dams have been constructed in the gullies using dry-stone walls, while in the valleys wind-breaks have been planted on the margins of the irrigated fields, which make up only 2 per cent of the total cultivated area but are highly productive.

Each SWC technique has evolved from a specific socio-economic situation. Forward-sloping terraces (talus) spread rapidly in the 1940s and 1950s, when extended families broke down into small nuclear families and the land was divided between brothers. At the same time the 1950s was a period of crisis, since the frontier with Algeria was closed owing to the war of independence and pressure on land increased greatly. Stone walls developed as soon as all good quality land was cultivated and people were forced to start cultivating more marginal land. The step terraces, as well as techniques for gully control, are directly linked to emigration since migrants' earnings permitted the use of new materials and hired labour.

Now, however, younger emigrants do not have the same attachment to the land and they tend to invest their earnings in other sectors, while those who have remained behind, and especially the young, are less inclined to spend their labour on the maintenance of SWC techniques. Tractors have become a much more common phenomenon and they are used even on steep slopes, with all the consequences for accelerating erosion that this entails. SWC activities in the valley bottoms will most likely be continued, because irrigated fields are the most productive and provide some security against drought. Step terraces are also being maintained, because absentee landowners plant these terraces with trees, which need little looking after and help to secure their land rights.

H El Abbassi

BOX 5.2: NEW CROPS AND MAINTAINED TERRACES BRING HANDSOME REWARDS

The Rif is a mountain range in Northern Morocco which relies on crops, livestock and woodland. Population density is high, with an average of six inhabitants per ha of cultivable land. The region is opening up to the outside world, with the development of infrastructure, growth of urban centres and the emergence of very profitable activities, such as growing *kif* (marijuana) and smuggling.

Agriculture in the Rif has been undergoing profound change. The increasing use of fertilizer is bringing higher yields, fallow is declining and new land is being cleared on steep slopes. Cereal growing, market gardening and tree cropping have traditionally provided the main sources of farm income, while animal husbandry is now in decline, with falling numbers of sheep and goats.

Kif growing has become a major illicit activity, especially for farmers in more outlying areas. This plant is well suited to the Rif Mountains and can be grown in rain-fed and irrigated conditions. Its yields are high and considerably more valuable than barley, 1 ha of *kif* produces an income of about 30,000 dirhams. Nowadays, some 200,000 people in the Rif make a living from this crop, which occupies more than 50 per cent of arable land in some regions. The high returns on this crop have encouraged farmers to expand the cultivated area, and to maintain and improve soil conservation structures. Stone bunds are the most commonly used technique for SWC, and are found particularly on irrigated land and parts of the old-established farm area. The use of conservation structures is most intensive where population density is highest. New structures are also being built on recently cleared land, given the new opportunities for earning profits from *kif*.

A Laouina

In the foothills of the Horst range, unlike the highlands, conservation measures are very recent and are found in only a few places. They tend to be introduced only where there is obvious erosion such as gullying, and farmers are primarily interested in making ploughing easier by eliminating cracks and gullies. These techniques are not used on the rest of the slope, even though it may be suffering serious loss of water and soil as a result of run-off. Some structures are maintained, while others nearby are not, for particular reasons which are not indicative of a general trend towards degradation. For instance, plots on which SWC structures are not maintained may belong to households which do not have the necessary family labour power, or to absentee landlords.

Advantages of traditional SWC techniques

Irrigated plots can produce two crops per year and yields are satisfactory when sufficient manure and irrigation water are available. Farmers recognize the usefulness of traditional SWC techniques, since yields on a plot where these are practised are much higher than a neighbouring plot which is unimproved. They recognize the advantages of the thick soil which builds up behind the walls and can be used for annual crops and tree planting at the edges where the build-up is deepest. Intensive cropping is possible thanks to these techniques. In the foothills, farmers appreciate embankments, since the only cereals to survive during dry periods are those planted just behind such structures.

The principles underlying SWC are still poorly understood by the farmers, even though they do use these techniques. For example, farmers build terraces and retaining walls on the lower part of the slope while destroying the vegetation cover further up, increasing run-off and causing water to overflow and destroy the structures. This shows the need to think beyond the level of individual plots and to take the whole landscape into account.

Heavy labour investment

Although the traditional techniques practised in the region are theoretically very expensive, the structures are constantly maintained and enlarged. They would never have been built if the farmers had tried to calculate the effort put into them. They are built by family labour at times when agricultural work is finished. At the moment, only one returned migrant has taken on paid labourers to maintain and enlarge the structures, at a rate of 35 dirhams for a day's work. This means that to construct 750m^2 of retaining walls on 1ha, an investment of more than 26,000 dirhams is necessary, which exceeds the market value of the land. No ordinary farmer can follow this example even if he is relatively well-off, so 'free' family labour is the only way to keep these techniques going.

The main constraints on the maintenance and expansion of conservation structures include fragmented land ownership and the distance between plots. This problem affects all farmers without exception. Land worked collectively, even by the members of a relatively large household, is often poorly maintained and it is only after such land is divided up into private holdings that individual owners pay more attention to their land. In fact, management of land and traditional techniques tend to be neglected at the end of each generation, when the head of the household becomes elderly and loses his power to manage. It is often the oldest son who takes over this responsibility when his father becomes infirm or is absent. Women have no specific role in managing land.

CONCLUSION

The oldest structures seen in the highlands may date back hundreds of years, yet they are still in place and some of them have never needed to be maintained or renovated. The soils built up behind the structures are thick and, therefore, represent a heritage to be safeguarded. The reason for building such structures is not the immediate gain derived from the improved field; rather, investment of labour is seen as something from which the family will benefit in the future, each generation marking its passing with additions to these SWC structures, such as by raising the height of old terraces or planting trees along new ones.

It is difficult to assess future prospects for the region given the rapid and profound changes that are now under way. In some areas, loss of vegetation cover is accelerating run-off from the slopes, destroying structures on the one hand and causing springs to dry up on the other. A new generation has emerged with other values and aspirations who see traditional ways of life as archaic and useless. Few young people would be able to construct terraces in accordance with the standards set by their forefathers, just as they would be unable to manufacture traditional agricultural implements such as wooden ploughs. However, the alternative of making a living in town is hard and has encouraged a return of some young people to the land, inflating land prices and demanding that a new balance in the rural environment should be found. Traditional techniques may continue to provide the basis for farming communities in the highlands and foothills, if higher yields and easier access to markets can be assured.

6

IMPROVED TRADITIONAL PLANTING PITS IN THE TAHOUA DEPARTMENT, NIGER

An example of rapid adoption by farmers

Abdou Hassan

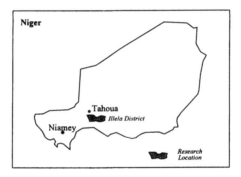

In 1988 the first phase of a soil and water conservation (SWC) project was initiated in the Tahoua Department in Niger, about 450km east of the capital of Niamey. This project was financed by the International Fund for Agricultural Development (IFAD) and concentrated its activities in the district of Illela. The past five years have witnessed a rapid spread of simple SWC techniques, which are particularly effective in rehabilitating degraded land. Farmers reap high yields from combining water-harvesting methods with manure and fertilizer use.

INTRODUCTION

The Illela district has an area of 7500km², an estimated population of 175,000 inhabitants in 1988 and is characterized by a landscape of fertile valleys and bare plateaux. Average rainfall in Illela was 448mm over the period 1950–89, but was only 368mm for 1980–87. The vegetation has suffered from the impact of drought and the pressure of people and livestock. Whereas the uplands had good vegetation cover about 40 years ago, they are now largely barren because, as population pressure in the valleys increased, people started to settle and farm on the plateau. Run-off from the uplands now causes significant damage to the fertile valley bottom lands. Since 1964, various SWC projects have tried to stop erosion, but none has had lasting impact, because land users did not maintain the conservation works. In the Badaguichiri Valley, which is the major valley in the Illela district, a major SWC project was implemented over the period 1972–81. Villagers were paid to construct stone bunds in the valley and the project used bulldozers to deep plough barren crusted soils. Little remains of this project and its activities.

A new approach: low cost and replicable techniques

A major objective of the IFAD-funded SWC project, by contrast, has been to introduce simple, low-cost techniques, which could easily be mastered by farmers. The project did not have earth-moving equipment or lorries, but occasionally hired them. Its staff was limited to the project director, an agronomist, to a deputy director, a specialist in soil and water conservation, and to ten extension agents. The main targets of the project were to construct contour stone bunds on 2300ha in four years and to develop 320ha with half moons (*demi-lunes*). In the first year of the project, it carried out activities similar to all other SWC projects which had intervened in this region. It hired a bulldozer for scarifying the land and constructed earth bunds (*fossés-ados*) which were partially carpeted with stones to stabilize the soil, while some contour stone bunds were also constructed. Food-for-work was used to remunerate the farmers.

The project changed course in its second year. Food-for-work was provided only in cases of serious food deficits and support shifted to tools-for-work and community infrastructure-for-work. This meant that villages that were particularly active in SWC could get a well dug or a school classroom built. Ten farmers were sent to the Yatenga region in Burkina Faso where they observed planting pits (Ouedraogo and Kaboré, this volume). Traditionally, planting pits were also used in the Illela district to rehabilitate degraded land, but only on a small scale and not very effectively. In the Yatenga farmers from Illela saw pits which were larger and into which manure was placed in the dry season. Upon return to their villages, the improved traditional planting pits, or *tassa*, were tried out on 4ha of land (see photo 4). In the village of Nadara in particular the impact of the improved pits was spectacular. As a result, 70ha of degraded land were rehabilitated using pits in 1990. Only in the pits was a reasonably good yield obtained in the drought of 1990. This convinced farmers of the great advantages of these techniques, so that in 1991 they treated 450–500ha, and in 1992 a further 1000ha. It is estimated that by mid-1995, about 6000ha of badly degraded land in the Illela district has been rehabilitated and this simple technique has also spread to adjacent districts. Although *tassa* were not mentioned in the original IFAD project document, they have now become the main focus of the project.

An emerging land market

Badly degraded land on the plateaux in Illela has become productive and a land market has now emerged. Farmers buy and sell degraded land for prices which increased considerably between 1992 and 1994. In the village of Batodi, 150,000 CFA Francs – the currency used within the West African Franc zone – (US$300) was paid in 1994 for a plot of degraded land, whereas a similar plot of land cost only US$120 in 1992. Prices depend on the availability of degraded land in a village. Some farmers have even started to sell their sandy soils in order to be able to buy degraded land which they will then rehabilitate. The emergence of a land market for degraded land shows that farmers believe that the *tassa* are an efficient and cost-effective tool for bringing degraded land back into production. This marketing of land may increase social inequality at village level, because it is likely that the poor will sell and the rich will buy.

Planting pits bring improved yields

Why are farmers so keenly adopting improved planting pits? The IFAD-funded SWC project is unique in the sense that it has measured the impact of *tassa*, half moons and contour stone bunds on a large number of farmers' fields; these fields are demonstration plots, but they are entirely farmer-managed. In the case of *tassa* and half moons, their impact was also measured on a number of experimental plots. On these plots, which have now been operational for one to three years, experiments have been carried out with different densities of *tassa* and half moons.

The results on the farmers' fields will be treated first here. The size of the demonstration plots is 400m^2. Over a period of four years, 470 farmers have been involved in such measurements and results have been obtained from a total of 932 plots. To be able to measure the impact of the conservation technique and of the use of manure and fertilizers, certain measurements were taken (see the table opposite above).

The table shows the way that farmers use planting pits to rehabilitate degraded land. Even in a year such as 1993 with low and irregular rainfall, where some areas received less than 300mm, an average yield of almost 400kg/ha was obtained. This is substantially higher than on untreated land. In years such as 1994 with good rainfall, the average yield was almost 1000kg/ha even without the use of fertilizers.

A comparison of the impact of different SWC techniques on millet yields in 1993 and 1994 (see the table opposite below) shows that in drought years half moons perform on average slightly better than *tassa*. This seems logical because half moons have a larger catchment area, so more run-off will be available to the plants. Contour stone bunds are outperformed by both *tassa* and half moons. In

Impact of improved planting pits (*tassa*), manure[1] and fertilizers[2] on millet yields

	1993[3] (kg/ha)	1994[4]
T0	144	296
T1	393	969
T2	659	1486

[1] Recommended quantity per *tassa* 0.5kg, average 5–6 tons/ha.
[2] Recommended quantity per *tassa* 50kg urea/ha, 125kg of phosphate/ha. Such quantities are only rarely applied by farmers.
[3] Low and irregular rainfall.
[4] Good rainfall.
T0: the situation without intervention; in this particular case the project either measured yields on fields without any SWC techniques or, if no yields were measured, it took the average yields of the district. However, this underestimates the impact of conservation and the use of manure and fertilizers because in the situation without intervention, yields are usually zero since nothing grows on barren degraded land.
T1: soil and water conservation technique + manure.
T2: soil and water conservation technique + manure + fertilizer.

Comparison of the impact of three SWC techniques on millet yields in 1993[1] and 1994[2]

	Tassa (kg/ha)	Half moons	Contour stone bunds
1993[1]			
T0	144	77	156
T1	393	416	292
T2	659	641	448
1994[2]			
T0	296	206	390
T1	969	912	671
T2	1486	1531	900

[1] Low and irregular rainfall.
[2] Good rainfall.
T0: The situation without intervention.
T1: Soil and water conservation technique + manure.
T2: Soil and water conservation technique + manure + fertilizer.

contrast, in a year of good rainfall, *tassa* do slightly better than half moons when only manure is used (T1) and both perform better than contour stone bunds.

The data show that the use of a SWC technique + manure (T1) has a very positive impact on millet yields; the use of a SWC technique + manure + fertilizer (T2) again improves the yield levels. However, in a year with low and erratic rainfall, the use of inorganic fertilizers may be an uneconomic proposition for farmers because the extra gain does not always cover the costs of fertilizer. The incremental yield from the addition of chemical fertilizer can be seen by subtracting T1 from T2. For *tassa* in 1993, a year of poor rainfall, T2 – T1 = 266kg/ha and for half moons T2 – T1 = 225kg/ha. However, in years with good rainfall, the effect of inorganic fertilizers is very much more positive and becomes financially worthwhile. For *tassa* in 1994, T2 – T1 = 527kg/ha and for half moons T2 – T1 = 619kg/ha.

The difference in performance between *tassa* and half moons is minimal, but it takes farmers more time to treat 1ha with half moons than with *tassa*. The IFAD project leaves the choice of technique entirely to farmers. In the standard layout, it takes 90–100 person days to treat 1ha with half moons. If these are dug by hired labour, the costs involved are about 30,000 CFA (US$60). It takes about 40 person days to rehabilitate 1ha with *tassa*. The costs of hired labour would be in the order of 12,500 CFA/ha (US$25). The difference in costs explains why farmers often prefer *tassa* to half moons. It also becomes clear why farmers are rehabilitating degraded land so enthusiastically with *tassa*. If they invest $25 per ha on hired labour, they will get a yield of about 400kg/ha of millet in a year with low rainfall (cash value US$80) and in a year with better rainfall about 1000kg/ha of millet (cash value US$160). The rehabilitation of barren and degraded land using *tassa* is an economic proposition, underlining why farmers and traders are increasingly interested in buying degraded land.

Experimental plot yields

On the experimental plots, three different densities of *tassa* and four levels of fertilization have been used. The table below shows that in 1993 the highest yields were obtained with the lowest density of *tassa*/ha. This is because 1993 was a year with low rainfall and a wider spacing of *tassa* meant that the

The impact of different densities of *tassa* on yields of millet in 1993 and 1994

(*tassa*/ha)	T0 (kg/ha) 1993[1]	T1	T2	T3
15,625	25	102	375	83
12,500	34	81	131	94
10,000	20	300	768	203
	1994[2]			
15,625	163	1883	1360	1609
12,500	188	1922	1534	1746
10,000	131	1242	1171	1414

[1] Low and irregular rainfall.
[2] Good rainfall.
T0: the situation without intervention.
T1: *tassa* + 0.5kg of manure.
T2: T1 + 50kg urea/ha + 50kg phosphate/ha.
T3: *tassa* only.

catchment area per *tassa* was higher, and hence more water was available to plants. In 1994, a year of good rainfall, the highest yield (1922kg/ha) was obtained with a density of 12,500 *tassa* and the use of manure only (T1). With manure and inorganic fertilizers, the yield of millet fell to 1543kg/ha. Even in the T3 situation (*tassa* only), the yield of millet was higher than T2 (1746kg/ha). The lower impact of inorganic fertilizers in this situation is due to the specific soil characteristics of the experimental plots.

A comparison of millet yields on the farmer-managed demonstration plots and the researcher-managed experimental plots shows that for T1 the yields on the farmers' plots are consistently higher than on the researcher-managed plots in a dry year. The opposite is the case in a year of good rainfall. The yield data indicate that there is still some scope for increasing yields on farmers' fields with a combination of *tassa* and manure only. Although more farmers are now using manure, the quality and quantity of the manure is still insufficient.

CONCLUSION

During its first phase, the IFAD-funded SWC project has concentrated its activities on the promotion of simple and low-cost conservation techniques. The rates of adoption, of *tassa* in particular, are astonishing. If the project closed now, the farmers would continue to rehabilitate degraded land; the contribution of the project is essentially to train the farmers and to support them by providing tools. Thousands of farmers have treated their own fields and by doing so they have considerably improved food security at family level.

7

REHABILITATING DEGRADED LAND

Zaï in the Djenné Circle of Mali

Joanne Wedum, Yaya Doumbia, Boubacar Sanogo, Gouro Dicko,
Oussoumana Cissé

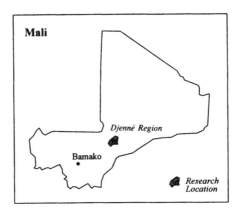

This chapter describes soil and water conservation (SWC) work being carried out by farmers in Central Mali. It focuses on the use of traditional planting pits, known as *zaï*, which have been adapted and improved, bringing higher and more secure crop yields, while also enabling formerly abandoned land to be brought back into cultivation. The NGO CARE has been supporting this work as part of the Agricultural Systems Project, in partnership with the Malian Association for Action Research and Development (AMRAD), in five districts within the Djenné Circle in the fifth administrative region of Mali.

INTRODUCTION TO THE AREA AND THE PEOPLE

The Djenné Circle has a total population of 129,500 inhabitants who live in 167 villages over an area of 4561km², giving an average density of 28 people per km². There are high levels of seasonal migration with 80 per cent of young men absent in some villages during the dry season. The rainy season extends from June to September, followed by a long dry season from October to May. The heaviest thunderstorms occur in July and August, while dry episodes are common in June and September and often in August as well, causing serious crop stress. The average annual rainfall is 538mm, which falls in 35 days.

The region is fairly flat overall, with occasional hillocks in places, and a variety of soil types:

- Laterites: very poor tropical ferruginous soils, damaged by erosion.
- Gravelly soils: very thin soils covered with fine gravel formed as the bedrock breaks up; these form the upper stratum in the uplands.
- Clays and clay–sand mixes: lowland areas which are flooded every year by the spate of the Bani and Niger rivers when climatic conditions are normal, and enriched by the alluvial deposits and sediment carried along by the rivers.
- Sandy soils: the intermediate stratum between the lowlands and uplands, characterized by the presence of vigorous natural regrowth, dominated by *Acacia albida*.

Patterns of land use

Under customary tenure, land is passed from parents to children, although in some cases village authorities may allocate land to other people who seek a place to farm. In formal legislative terms, all land belongs to the state and thus there is some uncertainty regarding the rights and responsibilities of local people, and their ability both to control access to their land and to resolve conflicts between different land users.

The basic food crops are rice, millet, sorghum, maize, cow peas and beans, supplemented by vegetables such as tomato, peppers, sweet potato, cassava, okra and lettuce grown on irrigated plots. The main cash crops are groundnuts and *dah* (*Hibiscus* spp) both of which have become more widespread since the great drought of 1973–4.

The major physical and environmental constraints to agriculture are inadequate and irregular rainfall and river spate, plant diseases, crop pests, lack of infrastructure and land degradation. Cropping practices have had to adapt to new climatic and economic circumstances. Low rainfall and the inadequate spate of the Bani and Niger rivers have led to a gradual transformation of a large part of the flooded area on which farmers traditionally grew rice into dry farming land where crops such as sorghum and millet are grown. Other changes related to market forces have brought about an expansion of groundnut and *dah* cropping.

In seeking to improve soil fertility, farmers rely above all on the natural regeneration of trees growing in the fields. *Acacia albida* is a species traditionally protected by farmers as it enriches the soil. Other trees such as shea (*Butyrospermum parkii*) and néré (*Parkia biglobosa*) are also protected because of their economic value and the shade they provide. Farmers in the area have considerable experience in protecting the young shoots of trees in the fields. Most development projects and programmes working in the region provide support to assist the regeneration of trees in places where vegetation cover is relatively sparse. To improve fertility, farmers also use leaf mould, compost and even small amounts of chemical fertilizer, while millet and sorghum stalks are left in the field after harvest.

The manuring of land destined for sorghum and millet growing is also made possible by means of contracts between farmers and herders. Some farmers dig a well in their field around which the herders camp during the hot, dry season, the animals grazing in and around the field by day are being penned by night.

Animal husbandry

Although this is primarily an agropastoral zone, the drought in 1983–84 caused a sharp fall in herd size. Some livestock keepers rely mainly on animal husbandry, while many agropastoralists concentrate on agriculture but keep a few head of draught oxen for ploughing. After the harvest, animals graze freely during the dry season and are joined by transhumant livestock from the north and east who pass several months on the stubble of millet, sorghum and ricefields, and in the extensive flood-plain pastures known locally as *bourgou*.

Livestock provide traction power and organic fertilizer for agriculture, which in turn supplies a proportion of fodder requirements in the form of crop residue. However, coexistence between crops and livestock is not without problems, because farmers no longer respect the traditional boundaries of agricultural land and so reduce the space available for grazing.

TRADITIONAL SOIL AND WATER CONSERVATION TECHNIQUES

Government and NGO projects are involved in improving traditional SWC techniques, such as *zaï*, stone lines, trash lines, live hedges, ploughing, ridging and mulching with crop residues. The subject of this study is the *zaï*, the most common traditional technique in the area whose use has been much encouraged by the currently dry climatic conditions. Over the last few decades, the formerly extensive flooded area has dried out completely, leaving bare, degraded land. The uplands have been affected by wind erosion, causing the abandonment of former cropping techniques and adoption by farmers of measures to conserve soils and moisture.

The *zaï* technique is only suitable for gravelly soils and clay slopes which generate high levels of run-off, and is used to rehabilitate degraded areas. The sites concerned are usually located within the family fields, on long-abandoned degraded land 'owned' by the village as a whole, or on the gravelly soils of the uplands and hillsides which, while forming part of village territory, had never been farmed before.

The traditional zaï

The zaï (literally, 'water pocket') is a technique used to develop degraded land for rainfed farming of millet, sorghum and maize. The traditional pits (*towalen* in Bambara) are of varying dimensions and size, usually about 20cm in diameter and about 10cm deep. Dug out with a traditional hoe, the earth excavated from the pocket is put outside the hole. As a rule, the pits are made in lines across the field, the farmer adding a small amount of manure to each pit before the first rains.

The improved zaï

The Djenné Agricultural Systems Project (SAD), whose objectives included the identification of simple techniques and technology which could be replicated at low cost, saw great potential in the traditional zaï technique. It therefore set about improving the technique in co-operation with several innovative farmers in Torokoro village. Improved zaï are a means of collecting run-off water within the field and are somewhat larger than the traditional pits, with holes about 30cm in diameter and 15cm deep (see photo 6). Several zaï are dug in staggered rows, following the contour lines but perpendicular to the slope, and the excavated earth is used to make a small half-moon bund downslope from each pit. Zaï may be associated with grass lines of *Andropogon gayanus*, euphorbia and *leptodonia* to combat erosion on steep slopes. The space between zaï varies from 0.8 to 1.00m, depending on the crop. As zaï are dug in degraded soils, they require considerable physical effort. Farmers can dig an average of 30–40 zaï in three hours, which means an investment of 600–800 hours per hectare.

Results over the last five agricultural seasons

Season	Number of participants	Area treated (ha)
1989–90	66	23.5
1990–91	110	36.4
1991–92	184	164.7
1992–93	653	489.8
1993–94	654	472.4*

*These figures only refer to the 17 villages taking part in the project in Djenné Circle and correspond to about 15 per cent of the cultivated uplands.

The SAD project, which began work in November 1988, had its first contact with the *zaï* technique in 1989 when it met Kassoum Fofana, who had pioneered the introduction of the technique in Torokoro. This innovative farmer had discovered the traditional *zaï* technique during a visit to relatives in Kouna village in the neighbouring Circle of Mopti. He decided to introduce the technique in his village after seeing the good millet and sorghum harvests obtained by his relatives despite irregular rainfall and the poor soils they farmed (see photo 7). Kassoum Fofana continued to experiment for two or three years before the arrival of the SAD project which suggested that he should collaborate with CARE International in Mali to try out improvements. The two main objectives of the project are to increase the cultivated area by rehabilitating degraded land, and to improve production and productivity on agricultural land.

Zaï techniques are currently among the most popular methods of developing uplands in the region. In using them, farmers themselves conduct experiments with the technique and crop selection. By way of example, for rapid recovery of very hard ground such as old termite mounds, Kassoum Fofana constructs circular bunds within which *zaï* are dug. An equally interesting experiment concerns the cultivation of varieties of rain-fed rice in *zaï* dug on the formerly flooded lowlands – ie, on clay slopes where such a combination produces a good sorghum crop after the rice harvest. Farmers have also been interested to compare the effect of different organic fertilizers used in the *zaï* (horse, sheep, goat and cattle dung, leaf mould, etc).

How the project helped

On arrival in the area, the project immediately set about identifying the traditional techniques used by the farmers, and subsequently helped farmers to develop these methods with an emphasis on training. After developing improved techniques, the project disseminated these in its zone of interventions through exchange visits and training sessions.

Advantages and impact

In order to assess the effect of the *zaï* technique, yields were compared on test plots treated with *zaï* and on traditionally ploughed control plots (see the table below).

Comparative results

Season	Crop	Yield *zaï* (kg/ha)	Yield ploughing (kg/ha)
1992–93	Sorghum	1494.4	397.2
1993–94	Sorghum	620–1,288*	280–320*

*Optimum sowing date.

Apart from the considerable increase in production, other benefits can be seen, such as improved spring flow, and more fodder and firewood, although the project has no statistics on these side benefits.

Most of the land on which *zaï* techniques have been applied was formerly unproductive, but it can now be sown. Moreover, according to the farmers, with both traditional and improved *zaï*, it is possible during the dry season to prepare the field ready for sowing at the onset of the rains, thus saving time in the rainy season. *Zaï* harvest and concentrate rainfall and runoff, and allow crops to survive spells of drought (up to 10–14 days) during the rainy season.

Constraints

The main drawback of *zaï* techniques seems to be that they require considerable physical effort, as they can only be dug with a traditional hoe.

In this area, all farmers, whether rich or poor, use SWC techniques, but only able-bodied people can dig *zaï* over a large area. Consequently, these techniques are a luxury for families who are short of labour unless they pay to use the services of the young people's association. Thus, labour shortages can be a constraint on the activity, as well as with the fact that some land is not suitable for these methods, such as sandy or clayey soils in the valley bottom and rice-growing areas. There are also problems with the transport and availability of manure for those farmers who have no carts and insufficient livestock.

Socio-economic considerations

So far, degraded land in the area has not been sold for the purpose of practising SWC techniques, unlike in central Niger where abandoned land now finds a ready buyer (Hassan, this volume). Nevertheless, digging *zaï* provides a source of income for some young people's associations (for instance, in Torokoro). Some women use *zaï* techniques, with nine taking part in this activity over the 1993–94 agricultural season. However, in most cases, as all fields belong to the family, women do not make the choice of which techniques to use. This matter is handled mainly by the men in this area, although the harvest is shared by the whole family.

Trends and recent history

Because of their effectiveness in increasing production, rehabilitating degraded land and assuring a harvest even in times of drought, *zaï* are becoming more and more widespread in the area. The good results achieved in the project villages have inspired surrounding villages to copy their example, so now abandoned land tends to be found only in sandy areas which are not suitable for the *zaï* technique.

CONCLUSION

SWC techniques such as *zaï* are becoming widespread in the area owing to low and erratic rainfall and land degradation. Farmers have little choice but to put efforts into rehabilitating their land, as agriculture is their mainstay. In these circumstances, *zaï* represent a viable option as they make it possible not only to rehabilitate degraded land which had been thought lost to agriculture but also to increase crop yields through the harvesting of rainwater. Some farmers even say that they have recovered all the degraded land in their villages by using the *zaï* technique. The effectiveness of the technique can be improved further by making more suitable tools, increasing the number of manure pits and compost heaps, and combining *zaï* with bunds, as some farmers are doing already.

8

A MEASURE FOR EVERY SITE

Traditional SWC techniques on the Dogon Plateau, Mali

Armand Kassogué, Mamadou Komota, Justin Sagara and Ferdinand Schutgens

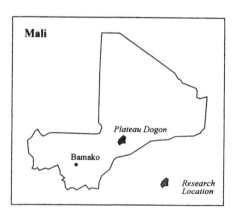

The Dogon Plateau is a classic case of a harsh environment, where people sought refuge from warring tribes. In order to survive on the plateau they had to make the most of scarce soils and erratic rainfall. The Dogon people are now famous, not only because of the detailed studies made by anthropologists of their customs and culture, but also because of the considerable range of soil and water conservation (SWC) techniques that they use. They have developed all kinds of stone lines and ridge systems to conserve soil and to make the best use of limited rainfall (see the figures on following pages). In particular, since the early 1950s the Dogon have developed a system of hand irrigation for the production of onions. A major feature of this system is that soil is transported by head and deposited on barren rock close to small dams. The soil is mixed with manure and on these newly created fields they produce yields of 30t/ha of

---

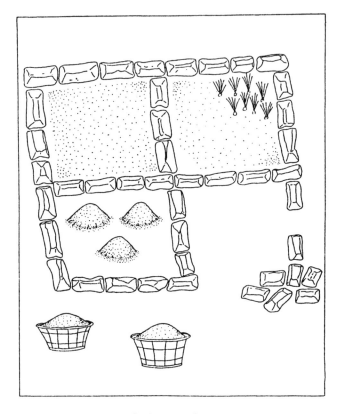

Onion gardens

onions. The total value of the onions and other irrigated crops was estimated to be in the order of CFA 2 billion in 1989 (US$ 4 million at current rates) and most of it is exported to other parts of Mali or to the Ivory Coast. Virtually every family on the Dogon Plateau uses SWC techniques, either on rain-fed or irrigated fields.

INTRODUCTION TO THE AREA AND THE PEOPLE

The Dogon Plateau, with a surface area of around 10,000km^2, is located in the central eastern part of Mali. It is a sandstone massif with an average altitude of 400–600m, its indented relief the result of long-standing erosion. To the east, large rocky outcrops are typical of the relatively uneven terrain. Narrow valleys several metres in depth are cut into the sandstone along the fracture lines. The central area is made up mainly of sloping plains on which sandstone outcrops

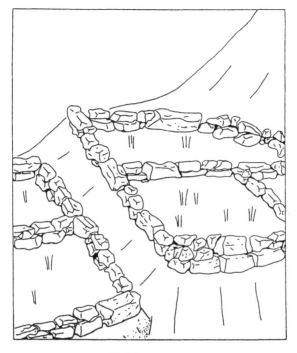

Hillside terraces

and soils of varying depth alternate, while the western part of the plateau is a plain studded with steep-sided rocks.

The whole plateau slopes down slightly towards the west, its water courses draining towards the Niger river. Only a narrow fringe of land along the cliff drains to the east. Water flow is seasonal, occurring for four to five months per year and, because of the importance of barren land, run-off ratios tend to be very high. Average rainfall is in the order of 500mm, but drought years occur frequently. The Sahelian tropical climate is characterized by a wet season from May to October and a dry season from November to May. There are heavy downpours in June and July, with more frequent but less intense rains in August, and there is extreme intra-regional variation in the spatial distribution of rainfall. The Dogon Plateau has not escaped the overall reduction in rainfall throughout the Sahelian zone in recent years, and has suffered from two very dry periods (1966–75 and 1981–90).

The two main types of soil are sandy–alluvial and sandy–clayey. Although the fertility of the sandy–clayey soils is low, they are still the most fertile in the region. They are degraded, the upper stratum having been eroded and the remaining subsoil forming a hard surface crust which is not very permeable. These soils lose a lot of water due to run-off.

Socio-economic context

The resident population was estimated at more than 200,000 in 1994, with a natural growth rate of 3.5 per cent per year. It is mainly made up of Dogon, with Peulh, Toucouleur, Sonraï, Mossi and other minorities. The cliff regions of the eastern Dogon Plateau are densely populated, sometimes with over 80 inhabitants per km^2, while the interior plateaux have an average density of 25 inhabitants per km^2. Out-migration, whether seasonal, semi-permanent or permanent, and movements within the area have been a feature of life in the Dogon Plateau for more than a century. Seasonal out-migration (August/September–May/June) mainly involves young single people in search of employment, but heads of family also move to the flood plains of the Niger river or towards southern Mali to harvest rice. The number of young girls leaving on migration is said to be rising dramatically. In years of serious drought, such as 1995, many Dogon leave the plateau and settle in parts of Southern Mali, a region with higher and more reliable rainfall (700–1300mm), a fairly low population density and good soils.

Land rights

Land belongs to the founder of the village. The rights of his patrilinear descendants over the land cannot be transferred to anyone. The head of the extended family, the oldest man alive, is the land chief who distributes collective fields, household fields and women's plots among his lineage. The land chief may lend fields and gardens to other lineages living in the same village or to the inhabitants of other villages who simply work the land. They usually owe only moral recognition, which they demonstrate to him during traditional Muslim festivals; these occasions also provide an opportunity to renew the loan or to break the contract.

A diverse range of crops

Apart from subsistence agriculture, farmers in the area are also involved in growing vegetables. In view of the unreliability of rainfed, staple food crops, the construction of dams and the keen interest in onions on the plateau, market gardening, which was virtually unknown until the early 1950s, is overtaking cereal growing in some areas. There is usually a correlation between the species grown and the cropping technique. By way of example, millet, which is the staple food, is grown in a system of earthen ridges (see the figure opposite), often in association with beans, whereas fonio, groundnuts and beans are only grown on flat land. In some parts of the Dogon Plateau, rice is also grown on restored or artificially created land downstream of uncultivated micro-catchments. Earth is transported and placed on sandstone slabs, where it is kept in place by stone bunds plugged with clayey soil to retain water (see photo 8). Hollows are dug around each calabash, aubergine, pepper and tomato plant, which farmers fill with manure and water. Such techniques are also used for sorghum and cash

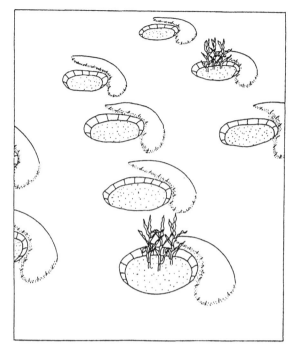

Planting pits or zaï

crops such as onions (see the figure on page 74), garlic and sweet potatoes (see photo 9). Tree seedlings (mango, guava, lemon, papaya, apple, banana, etc) are usually produced in nurseries. When planting out, farmers make a hollow around the seedlings which they fill with manure and then water the plant until it is able to survive on the natural water supply.

Insufficient rainfall has led to a gradual loss of interest in fonio, a decline in sorghum in favour of millet, and a fall in rice in favour of sorghum. The growing demand for certain agricultural products in the market has also influenced crop selection, with onions, groundnuts and tomatoes gaining ground, while cultivation of dah (*Hibiscus* spp) has expanded greatly over the last five years.

The role of trees

Farmers recognize the merits of particular trees in their fields. Among trees with beneficial effects, *Acacia albida* is always top of the list, as farmers believe that its roots contribute nitrogen and soil rich in nitrogen quickly absorbs rainwater. The leaves of *Combretum glutinosum*, *Lannea microcarpa* and *Guiera senegalensis* enrich the soil, while the remains of dead baobab and tamarind trees are believed to have enriching properties.

Mound-making in fields

Livestock

The Dogon are usually agropastoralists, using livestock dung to enrich agricultural land. The number of cattle has decreased considerably in recent years, mainly due to poor climatic conditions, while donkeys, which play a major part in transport, have increased in number.

Constraints

Constraints on agriculture derive from both natural and social causes. Inadequate rainfall, the lack of arable land, insufficient quantities of manure and soil impoverishment are combined with constraints relating to land rights. Only landowners have full rights over land. Those who have no land can only apply for it from those who do, the latter are free to refuse, and applicants usually have no choice about the quantity or quality of the plots nor the duration of the loan. Borrowers are not free to do what they like on the plot (eg, plant trees). Ownership of land is a very sensitive issue which sometimes constitutes a major problem between villages where, for example, one settlement is installed on land which belongs to another.

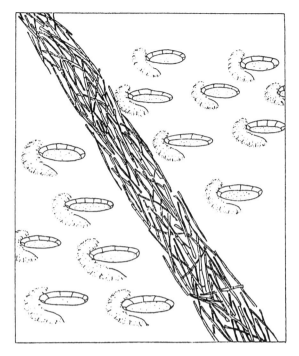

Trash line of millet stalks and zaï

TRADITIONAL SWC TECHNIQUES

All farmers on the Dogon Plateau are interested in SWC, owing to the need to conserve soil and water in this harsh environment and owing also to land shortages arising from demographic growth. Traditional SWC techniques apparently date back to before the arrival of the current occupants of the Dogon Plateau. The practices are said to have evolved from the intervention of supernatural creatures who used to 'drive' stones to the site where the lines and bunds were to be constructed. Men and women, poor and rich, practise SWC. However, the rich may do more than the poor in quantitative terms if they can call on groups of young people in the village. Payment varies depending on the village, the size of the group and the type of assistance required, the amount received being paid into the group or village fund and used for community projects.

Techniques adapted to the soil

Traditional techniques are adapted to the type of soil, its physical character-istics providing a set of constraints to be tackled by specific conservation measures. Sandy soils are characterized by ease of infiltration, and on these, trash lines are most commonly used, as water erosion is not usually too much of a problem. Gravelly soils are compact and hard, infiltration is limited and tillage with a hoe is difficult. *Zaï*, ridges in the form of squares or honeycombs, and stone lines and bunds are most commonly used in such cases. Clayey soils are heavy, difficult to work in times of drought, and infiltration is poor, but the puddles which form on the surface can be used for sorghum and rice cropping. The most common SWC technique for clay soils is the stone bund. On stony soils, farmers can make small-scale improvements by moving stones to the edge of the plot to obtain a larger growing area. When seeking to restore a barren, degraded field to productivity, the farmers usually combine *zaï* and earthen ridges.

The range of traditional techniques practised by the Dogon are shown in the table opposite. Dogon farmers often combine several techniques. Increasing population pressure and deteriorating natural conditions make it essential to use SWC techniques in almost all fields. Nevertheless, some techniques are becoming less common, such as terracing in the hills as practised by the first settlers who sought shelter from invaders. These days, peaceful conditions make it easier for people to exploit the extensive plains lying below the plateau.

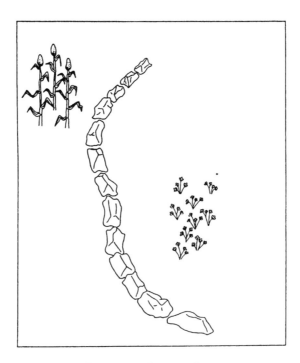

Traditional stone line

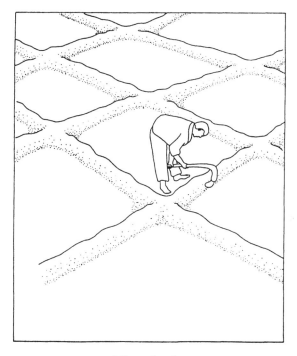

Micro-basins

Description of Traditional Techniques

Type	Form and site	Main purpose
Stone walls	Steep rocky hills, gravelly slopes and stream beds, low walls	Retains earth and holds back water
Stone lines and bunds	Along the contour, row of stones reinforced with earth	Against sheet erosion and gullies
Micro-basins	In fields, in squares or honeycomb pattern, each side 0.5–2m long	Captures rain falling within field and collects around plants
Mounding	In fields, small conical mounds formed during weeding	Conserves moisture and provides plant matter
Trash lines	Millet and sorghum stalks, and branches of trees, along contours	Against gully and sheet erosion
Grass barriers	Along gullies and ravines grass establishment encouraged	Protects from erosion
Creating fields	Grid of stones filled with earth built on bare rock, next to water source	Plots for vegetable gardening with easy access to water for irrigation

Constraints on SWC

There are two main constraints to SWC on the Dogon Plateau – labour shortages and land tenure. SWC work suffers from labour constraints because it is usually conducted between January and June, a period which coincides on the one hand with intensive onion cultivation and harvesting and, on the other, with the out-migration of the most able-bodied villagers. With regard to land tenure, it has been observed regularly that farmers who borrow land and treat this land with SWC techniques usually lose the benefit from such investments because the owner quickly withdraws the land for his own use.

Role of women

Women are active in various SWC activities. While weeding millet, they build up a network of earthen ridges (micro-basins) which criss-cross the fields. They take part in putting manure in the *zaï*, the digging of which is a man's job, and although women do not usually have fields of their own unless they are widows, stone bunds can be seen on small plots allocated to them by their husbands.

Extension and improvement of SWC techniques

Many projects and government services have undertaken SWC activities on the Dogon Plateau, with several attempts to improve traditional SWC techniques. For instance, a major German-funded Agricultural Extension Project (Projet de Vulgarisation Agricole sur le Plateau Dogon), which began in 1985, favoured the following modifications:

- Constructing stone bunds on contour lines determined by means of a water level or 'A' frame.
- Use of small stones in making bunds.
- Identifying the proper spacing of bunds.
- Placing the foundation stones of the bund in a shallow trench.
- Making stone bunds wider.
- Improving the traditional planting pits or *zaï*.

This project has supplied equipment to 97 villages, while the village reafforestation projects run by the Forestry Service and the National Agricultural Extension Programme have equipped 5 and 34 villages respectively.

The Yawakanda Experimental Erosion Control Techniques Project, launched by the Malian NGO HDS (Harmonious Development in the Sahel) brings together four partners in Bandiagara: HDS, the Forestry Service, French volunteers and the German-funded Agricultural Extension Project. This project is designed to harmonize interventions in the field of SWC. Training is given to a team of farmers (both men and women) in each village who then train other farmers and check that SWC work is done properly. Contour stone bunds have been constructed through community labour, thus protecting land in the process of degradation and restoring already degraded land. Material support (hammers, wheelbarrows, crowbars, etc) has given a boost to SWC efforts. Improvements in stone bund construction have made it possible to recover areas affected by gullies or serious sheet erosion, which would not be easy to rehabilitate by means of traditional techniques.

Farmers on the plateau are now keener on improved techniques following initial hesitations due to the fact that:

- Tracing the contour lines, which is sometimes rather difficult, was seen as a waste of time.
- Some techniques take more time than traditional practices.
- The objectives set by project technicians and the farmers' expectations are not always the same.

CONCLUSION

The Dogon Plateau demonstrates well how farmers have had to develop the means of carving a living from difficult circumstances. Shortage of land and fertile soils have obliged the Dogon to construct a variety of techniques for capturing rainfall and using whatever resources are available locally to assure a harvest. In some parts of the plateau, indigenous techniques continue to be maintained and expanded, and much ingenuity has been shown in the creation of plots of land for cultivation of high-value vegetables, such as onions. Elsewhere, traditional SWC methods are being abandoned because they are not effective enough in coping with accelerated erosion and run-off. Growing population pressure and regular droughts are inducing many Dogon to migrate down to the plains and to the better watered regions of southern Mali. This migration of the workforce may put in jeopardy the labour-intensive systems of SWC and crop production which have supported the Dogon people for generations.

9

THE *ZAÏ*: A TRADITIONAL TECHNIQUE FOR THE REHABILITATION OF DEGRADED LAND IN THE YATENGA, BURKINA FASO

Matthieu Ouedraogo and Vincent Kaboré

Since the early 1980s, improved traditional planting pits or *zaï* (in Moré) have been rapidly adopted by farmers in the Yatenga region of Burkina Faso. Using *zaï*, thousands of hectares of degraded land have been brought back to productivity. It is striking that while in other regions of the Central Plateau of Burkina Faso this relatively simple technique has hardly been adopted, in certain regions of Mali and Niger the *zaï* are rapidly becoming popular (see Hassan, this volume). Although some projects in the Yatenga region have been actively promoting the use of *zaï*, most farmers who adopt this technique do not receive any project support. This case study describes the major characteristics of the *zaï* and analyses their advantages and disadvantages.

INTRODUCTION TO THE REGION

The long-term average rainfall of Ouahigouya, the capital of the Yatenga region, is well over 700mm, but the average for 1973–83 is only 562mm, with rainfall in 1984, 1985 and in 1990 well below these figures. The Yatenga region of Burkina Faso has some of the highest population densities in the country, with central areas supporting more than 100 people per km². Until the end of the 1970s, these high population densities did not result in agricultural intensification. The common reaction among farmers was to expand agriculture on to land which was very marginal and to migrate temporarily or permanently to the Ivory Coast. Many families in the Yatenga region now have relatives who grow coffee on smallholdings in the Ivory Coast. However, since the improvement of the *zaï* and the introduction of contour stone bunds in the early 1980s, farmers have started to intensify land use. The introduction of these improved traditional soil and water conservation (SWC) techniques was also accompanied by a systematic adoption of compost pits. Efforts have also been made to enclose livestock in order to reduce grazing pressure, to allow the natural regeneration of the vegetation and to produce more and better quality manure. Unfortunately, efforts to enclose livestock have not been successful because it requires too much work collecting and transporting fodder, in particular by women. However, since the 50 per cent devaluation of the CFA Franc, used in much of the Sahel in January 1994, meat produced in the Sahel can compete again in the market of Abidjan with that imported from Argentina and the European Union. As a result, the prices of goats and sheep have increased and the enclosure of small stock has become more attractive. It is clear that the introduction of technological changes helped to initiate a process of agricultural intensification in the Yatenga region, but macro-economic policies such as devaluation also influence significantly the degree and pattern of the intensification of agriculture and livestock at regional level.

ZAÏ: A DESCRIPTION

The *zaï* usually have a diameter of 20–30cm and a depth of 10–15cm (Wright, 1982), their dimensions varying according to the types of soil on which they are dug. They tend to be larger on lateritic soils, which have a low capacity for water retention, than on soils which are less permeable. The number of *zaï* per ha can vary from 12,000 to 25,000 and depends on their spacing. When the spacing is 80cm, the number of *zaï* per ha is about 15,000. The larger the planting pits and the bigger the spacing, the more water can be harvested from the uncultivated micro-catchments. The excavated earth is put downslope and, if done properly, it adds to the capacity of the pits to retain water. *Zaï* are used mainly to rehabilitate lateritic soils as well as what the Mossi call *zipelle* – ie, sterile, crusted land, with a hard-pan surface. These soils are rock hard and neither rainfall and run-off, nor plant roots can penetrate the crust.

The pits are dug during the dry season from November until May (see photo 10). This is one reason why *zaï* have been readily accepted by the farmers, and why a system of tied ridges developed and promoted by researchers has been rejected, because they must be constructed during the rainy season. The digging of 1ha of *zaï* takes about 60 work-days (average five hours per day). During the dry season, the *zaï* trap litter and fine sand deposited by the wind. Farmers add some manure to the pits, which attracts termites which dig channels, allowing water to infiltrate deeply into the soil. Not only do they increase the porosity and the water-holding capacity of the soils, but they also transport nutrients from deeper layers to the top and the other way around. The termites have become major allies of the farmers in their struggle for the rehabilitation of degraded land (Critchley *et al*, 1994). The *zaï* are effective because they concentrate water and manure in the same spot. Some farmers have already sown in the *zaï* before the rains arrive. This means that the seeds profit from the first rain.

Only indicative data on yields are available for the Yatenga where, unlike in Niger, the impact of *zaï* on yields has not been measured systematically (Hassan, this volume). In a year of average and well-distributed rainfall, yields are obtained which vary from 500 to 1000kg/ha of millet or sorghum. The biomass production varies from 2 to 4 tons of millet or sorghum stalks/ha, depending on rainfall and soil fertility (see photo 11).

In the second year, farmers sow into the existing hole or, if the spacing of the *zaï* is large, they may dig new ones in between the existing ones. After five years or so, the entire surface may have been improved by the *zaï* and termite action (Roose *et al*, 1994).

Manure added to the pits by farmers contains grass and tree seeds which have already passed through the digestive tract of cattle, goats and sheep, so many seeds germinate readily. The seedlings benefit from the concentration of water and nutrients in the pits and show considerable growth rates. Roose *et al* (1994) have identified 13 species of trees and shrubs, and 23 grass species on a field rehabilitated with *zaï*, and many farmers in the Yatenga now protect with great care the trees they have established on rehabilitated land.

Origins of the *zaï*

Planting pits or *zaï* were used in the Yatenga region before the 1980s but only on a very small scale. They were largely abandoned in the 1950s and 1960s because this period was characterized by above-average rainfall, so *zaï* were not needed. Recurrent drought in the late 1960s and early 1970s provoked misery, to which the improvement of the *zaï* was a reaction. The technique was improved around 1980 by Yacouba Sawadogo, a farmer from the village of Gourga, 5km east of Ouahigouya (see photo 12). Several other farmers also claim the paternity of the improved *zaï*.

Adoption and diffusion

Projects like the OXFAM-funded 'Project Agro-Forestier' and the German-funded 'Projet Agro-Ecologie' have systematically promoted the use of *zaï*, but in many cases farmers have spontaneously adopted *zaï* just because they saw them applied by others in their fields. It is striking that *zaï* are used mainly in the Yatenga and hardly at all in other parts of the Central Plateau, such as the provinces of Passoré, Bam and Sanmatenga, which also have locally high population densities, as well as vast tracts of barren crusted soils which could be rehabilitated. Recently, *zaï* have started to be applied on a small scale in the region of Pissila (Sanmatenga province) as a result of farmer-to-farmer visits. Such visits are increasingly considered to be the best tool for the diffusion of agricultural and institutional innovations. One of the most significant examples is the adoption and diffusion of improved traditional planting pits in Niger as a result of a farmer's visit to the Yatenga in 1989 (Hassan, this volume).

Advantages

The *zaï* have many advantages, most of which are readily summed up by the farmers:

- They concentrate rainfall and run-off, and for that reason crops are less susceptible to dry periods within the rainy season. They can tide a crop over a dry spell, which also saves seeds and labour.
- *Zaï* concentrate manure and are therefore a means of economizing on its use. This is particularly attractive to farmers with few livestock.
- They capture wind-blown soil and litter, thereby improving soil conditions.
- *Zaï* allow the reintroduction of soil fauna (termites, etc), which improves the soil structure.
- They can be dug gradually during the dry season so that when the rains arrive, the fields are already prepared.
- Seeds and young plants are protected against gusts of wind as well as against run-off at the beginning of the rainy season.
- *Zaï* permit the rehabilitation of strongly degraded land which is of vital importance in a region characterized by high population densities, considerable pressure on scarce land and serious land degradation.
- They make it possible to get a yield, even in the first year, which is generally higher than the yields obtained on fields already under cultivation.
- It is possible, particularly in the first few years, to economize on labour for weeding, since weeds do not grow on the land between the *zaï* where the crust is not broken.
- They stimulate the regeneration of woody vegetation.
- They contribute locally to a replenishment of the groundwater table.
- It is a simple technique which can be mastered by all farmers.

Disadvantages

The advantages clearly outweigh the disadvantages, otherwise the farmers would not voluntarily adopt *zaï*, but there are a few important disadvantages:

- They require a considerable investment of labour which can be a problem for certain families.
- It is not possible to mechanize the digging of *zaï*. Roose *et al* (1992) tested an ox-drawn implement with one tooth for subsoiling every 80cm (*sous-solage croisé*), which could halve the time needed for digging *zaï*. The problem is to get oxen which are sufficiently strong to do the subsoiling.
- It is not possible to use a plough on land where *zaï* have been dug.
- It is often advisable to combine *zaï* with the construction of contour stone bunds because the bunds protect the *zaï* against strong run-off, but this adds further to the already high labour requirements.
- Where soils are already shallow, they become even shallower when *zaï* are dug. In those cases, farmers should not plant in the pit but rather on top of the excavated soil in order to maximize rooting depth.

CONCLUSION

Zaï make an important contribution to food security in the Yatenga region by assuring that some harvest can be reaped even in a year of low rainfall. Wealthier farmers are more likely to benefit from *zaï* than relatively poor farmers, because the former can invest more time and money in getting *zaï* dug, and they usually have access to larger supplies of manure. This study found that poor farmers also seem to gain considerable returns from this simple technique. Although their land is usually marginal, they often have a few sheep and goats, which means that a little manure is available for fertilizing their fields. *Zaï* are ideal for marginal land where a hard pan surface promotes run-off into the pits. Furthermore, they are economical in the use of manure and thus allow poorer farmers to make the best use of their meagre resources.

1. *Traditional terraces in the Atlas mountains, Morocco.*

2. *Traditional terraces in the central Rif mountains, Morocco.*

3. *Stone terraces along the river bank create narrow plots for cultivation, Morocco.*

4. *Newly dug* tassa, *prepared at the end of the rains when soils are still easy to work. Tahoua, Niger.*

5. *Half-moon basins rehabilitate degraded land and support a crop of millet, Niger.*

6. *Improved traditional planting pits in the Djenné region, Mali.*

7. Mr Fofana describes his experiments with improved planting pits in the Djenné region of Mali.

8. Rice cultivation in fields built on rock, Dogon Plateau, Mali.

9. *Hand-irrigated onions and peppers grown in small Micro-catchments constructed on a bare rock, Dogon Plateau, Mali.*

10. *Digging* zaï *on Mr Sawadogo's field, Burkina Faso.*

11. A good crop of sorghum on a rehabilitated field, Burkina Faso.

12. Mr Sawadogo in a field rehabilitated by zaï, *Burkina Faso.*

13. *Transport of grass for mulching, Burkina Faso.*

14. *Women mulching fields in Tagalla, Burkina Faso.*

15. *Small pool among the dunes, with* massakwa, *near Maiduguri, northeastern Nigeria.*

16. *Tied ridges shortly after rain, Zimbabwe.*

10

MULCHING ON THE CENTRAL PLATEAU OF BURKINA FASO

Widespread and well adapted to farmers' means

Maja Slingerland and Mouga Masdewel

Most soils in the semi-arid Sudano-Sahelian zone of West Africa are liable to crust formation which produces considerable run-off during heavy rainstorms. The run-off water selectively washes away any organic matter and nutrients from the surface layers. Thus, not only water but also soil fertility are lost. Several traditional soil and water conservation (SWC) techniques to fight this phenomenon can be found in this region (Roose, 1988). Mulching, which is one of them, will be treated in detail here. Mulching is examined in the province of Sanmatenga and the village of Tagalla.

INTRODUCTION TO THE AREA

In the densely populated Central Plateau of Burkina Faso, most fields are permanently cultivated. Demographic pressure is increasing and, with the decline in opportunities for labour migration to the Ivory Coast, some migrants have returned and settled in their villages again. In this region few crop residues remain on the field after the harvest as they are used as feed for livestock or as fuel. The bare soils, largely loamy in texture and lacking any protection, are thus susceptible to crust formation, run-off and soil erosion. Traditionally, farmers put a mulch on small parts of barren land with a hard crust as a means of rehabilitating it, as well as spreading it on cultivated fields on which the yields had declined. Until the early 1990s, this technique continued to be practised only on a minor scale because farmers felt that it involved considerable work cutting grass, transporting it to the fields or transporting millet stalks and spreading them. However, since 1992, the practice of mulching has spread rapidly. In the months just before the rainy season (from February to May) many people, men and women, old and young, can be seen cutting grass and transporting it in heaps on their heads or by donkey cart to the fields (see photo 13). Cases have been found where young migrants who have returned from the Ivory Coast were allocated marginal land and made an enormous effort to improve the quality of these fields through mulching. Elsewhere, old women mulch their small plots (*beolgo*) on which they largely depend. Where grass is not sufficient, people collect dry leaves of various trees, in particular of *Butyrospermum parkii* or sheanut. A light form of mulching is used by Mossi farmers. This method consists of covering the soil with a layer of about 2cm of dry grass, equivalent to 3–6t/ha. The advantage of mulching is not only the fertilization of soils by decomposition, but also the attraction of termites. Termites create passages in the soil, thereby destroying the crust, increasing soil porosity and permeability. Termites also stir and mix large amounts of soil (Mando *et al*, 1993). All these factors together create more favourable conditions for the development of roots. The remainder of the mulch, after termite consumption, is sufficient to absorb some rainfall and to decrease run-off (Roose, 1989).

The province of Sanmatenga

Tagalla village is situated in the province of Sanmatenga. A farming systems study for this province was recently carried out which showed that mulching was only a sideline, but the results provide an idea of its relative importance (Barning and Dambré, 1994). Data are based on a sample taken in three villages in the province, one in the north (Kogyende) and two in the south (Dembila and Ouanané). The surfaces mulched in these villages and by the two main ethnic groups are presented in the table opposite which shows that in the north, 74 per cent of the households mulch part of their fields, whereas this is 43 per cent in the south (Ouanané).

The study also found that the use of animal traction and organic fertilizer was much more common in the south than in the north and, as might be expected, farmers using animal traction and organic fertilizer were less involved in mulching. On the smallest farms (0–4ha), 39 per cent of farmers put a mulch on their fields, but on the larger farms (4–8ha), 64 per cent mulch part of their land.

Percentage of household fields mulched for different villages and ethnic groups

	Villages			Ethnic Groups		
	Kogyende	Denbila	Ouenané	Mossi	Peulh	Total
Number of households	101	48	72	150	71	221
Surface mulched (per cent)						
None	25.7	81.3	56.9	33.3	74.6	48.0
0–25 per cent	61.4	14.6	30.6	50.0	22.5	41.2
25–50 per cent	11.9	4.2	11.1	13.3	2.8	10.0
50–75 per cent	1.0	–	–	0.7	–	0.5

Larger farms have larger bush fields and these are more often mulched than fields closer to the homesteads which tend to receive more manure. Farmers on larger farms also often have a donkey cart which facilitates the transport of grass. The study also revealed that whereas the majority of the Mossi (67 per cent) mulch their land, only a minority of the Peulh (25 per cent) do so. This can be explained by the fact that the Peulh are agropastoralists who have larger numbers of livestock and therefore more manure, while their fields are smaller than those of the Mossi, who depend more on farming. A positive relationship was also found between the total area mulched and investment by farmers in SWC. Finally, the table below shows a positive relationship between available work-force per ha and the area mulched. This suggests that labour may be a constraint on the area which can be treated in this way.

Tagalla

in Tagalla, a village on the Central Plateau located in the western part of Sanmatenga province, a more in-depth study on mulching was done with 49 households practising this technique. The farmers distinguish four main soil types in their village: *bolé* (clay), *baongo* (valley bottoms), *zipellé* (sandy–loamy soils with a hard crust) and *zegdega* (lateritic soils). All 49 farmers claim that the *zipellé* suffer most from erosion and run-off, and acknowledge that the eroding

Number of active household members per ha cultivated in relation to the percentage of the surface mulched

Surface mulched (per cent)	Active worker per hectare	Number of households
None	1.71	105
0–25	2.11	91
25–50	2.60	22
50–75	2.67	1
Total	1.96	219

factors are: rain, wind and human action. They mention that erosion is accelerated by a decrease in the vegetation cover, especially the progressive disappearance of trees (50 per cent) which is said to result in bare and hot soils (6 per cent), a decrease in rainfall (10 per cent) and drier soils (10 per cent). This results, according to the farmers, in a decrease in crop production and in weight loss for cattle. The farmers generally know of several methods to fight erosion. Stone bunding and mulching are mentioned by 36 per cent and 38 per cent of the farmers respectively, while 25 per cent mentioned grass strips. Mulching and grass strips are traditional practices, whereas stone bunding was introduced in the 1980s. The main constraint for mulching is the scarcity of grass and the distance over which to transport it. Only 12 per cent of the farmers claimed to be able to find grasses nearby; the others considered them a long way from their fields. However, grass is light and small quantities can be carried even by old women. Most transport of grasses is done by head, with only 10 per cent of the farmers using donkey carts.

Loudetia togoenses is the only species of grass used for mulching. It can be found in reasonably large quantities as homogeneous layers on very poor soils on nearby hills. Its alternative use as fodder ends as soon as it starts to flower because then it grows spikes which discourage livestock from eating it. Other grasses are not available in the same quantities, either because they are eaten by livestock or because they are cut and used at household level for handicraft, thatching or fodder. The main period for cutting and mulching is May, although some farmers may already make a start in February (see photo 14). In this period the grass is totally dry and unfit for livestock consumption. Farmers want to mulch all their fields but are constrained by the time required. In the majority of cases, two weeks of work is put into mulching, including cutting and transport. With this investment in time, none of the farmers managed to cover all their fields. The criteria used by farmers to choose the fields to be mulched are the following: low soil fertility (47 per cent), prevention of soil degradation (23 per cent), maintaining or increasing soil humidity (17 per cent) and a decrease in crop production in general (10 per cent). The benefits of mulching they mention are an increase in crop production (36 per cent), maintaining and increasing soil moisture (30 per cent), an increase in soil fertility (23 per cent) and protection of the soil against wind, rain and the sun's heat (5 per cent). The main crops which benefit from mulching are sorghum and cowpea which are mostly intercropped. Sorghum is the staple food of the area. Apart from mulching, some farmers also use manure (42 per cent) or even inorganic fertilizers (16 per cent) on the same fields. Some farmers (23 per cent) burn their mulch shortly after the beginning of the rainy season, partly to facilitate ploughing at the start of the planting season and partly to facilitate weeding afterwards.

Although farmers know the advantages of mulching, they are not interested in applying mulch to fields other than to those currently being cultivated. Forty per cent of farmers also claimed that they have no land left in fallow. For many farmers, stone bunds and mulching are complementary measures, because once the investment is made in stone bunding, mulching brings added benefits, thereby increasing the return on investment in conservation works.

CONCLUSION

Mulching is a traditional soil- and water-conservation technique which is now widespread and well adapted to farmers' means. However, many questions remain to be answered. What triggered the rapid adoption of mulching by thousands of farmers on the northern part of the Central Plateau, since 1992 in particular? It cannot be due to the activities of development projects or national programmes because none made a specific effort in the pre-1992 period to promote mulching. It may have been triggered by the low rainfall received in 1991. In 1995, by contrast, following a year of high rainfall, the area mulched was considerably less than the previous year. But low rainfall cannot be the single reason triggering mulching because there were years of low rainfall in the 1980s which did not lead to such an increase in mulching. In the 1980s, the government enacted a number of measures aimed at reducing environmental degradation, such as the campaign against bush-fires. As a result, such fires have become a rare phenomenon, which has led to an extension of grass cover in this region. Further investigation is needed of other issues, such as how villagers manage the areas where *Loudetia* grasses grow. Do they control access to grass, or can every villager cut as much as he or she can handle? In some villages, those who cause grass- or bush-fires are now sanctioned and have to pay a fine, which indicates a growing interest and concern among villagers for ensuring the maximum availability of materials for mulching their land.

11

FIRKI-MASAKWA CULTIVATION IN BORNO, NORTH-EAST NIGERIA

Are Kolawole, J K Adewumi and P E Odo

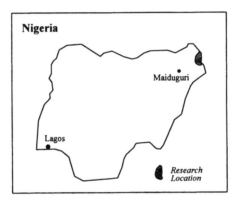

The collapse of the South Chad Irrigation Project in north-east Nigeria has led to a resurgence in indigenous soil and water conservation (SWC) practices, accompanied by dramatic shifts in the crops cultivated. Farmers are now reverting back to the more sustainable production of local staple crops such as *masakwa* sorghum.

INTRODUCTION TO THE PEOPLE AND THE AREA

Borno in north-east Nigeria is a region of relatively flat, sandy, undulating plains which slope gradually towards Lake Chad. The plains are bordered on the south and south-west by the Bama beach ridge, on the east by the Republic of Cameroon and on the north-east by Lake Chad. With a land area of 69,435 square kilometres, it is sparsely populated, with an average population density of only 37 people per km².

Situated within the Sudano-Sahelian zone, the area has a long dry season from September to May, and a relatively short wet season from June to September. Rainfall is low and highly variable, diminishing towards Lake Chad, where, on average, 300mm falls during the wet season. The rainfall varies from between 250mm in the north of Borno state to over 500mm in the south. The low level and unpredictable pattern of rainfall make agriculture in this region a highly precarious activity.

Most of the people in Borno belong to the Kanuri ethnic group, where customary land ownership is obtained either by clearing land with the permission of the village head, by inheriting a plot, or by being given a plot by a previous owner. The owner of a plot of land is seen as a mere trustee who has no right to sell the land. Ownership also depends on the continued cultivation of the land; if it is left to lie fallow for too long, then the land is given back to the group. Cohen (1967) has identified four different types of land ownership around villages in Borno:

- Land owned outright by the household head.
- Land which has been cleared by the head of the household which may become his own after long years of use.
- Resting land, a farm plot that has been in use but whose depleted soil the owner will probably use again in the future.
- Free unused land which can be used by newcomers or those trying to expand their farms.

Masakwa cultivation

Origin and evolution
By tradition farmers, the Kanuri are known to have cultivated hardy, drought- and pest-resistant crop varieties in response to their inhospitable environment for quite some time. A European explorer in the 17th century identified a number of hardy, early maturing and high-yielding varieties of sorghum in use (Nachtigal, 1980). In particular, *masakwa* (*Sorghum bicolor*) was identified as a crop which is grown across the entire Sahelian zone.

Masakwa is a drought- and cold-tolerant short-day sorghum grown specifically on the fertile *firki* clay soils around Lake Chad, which retain moisture during the dry, cool period from September to January. *Firki* soils are inherently fertile and have the potential to store adequate water for crop growth during the wet and dry seasons to which *masakwa* is ideally suited. Earth bunds further enhance the water-retention capacity of the vertisols, which are often built across the gradient of the gently sloping land to trap any rain or floodwater.

The Kanuri have also developed a number of other farming techniques in response to rainfall, the extent and depth of flooding, soil conditions and the reaction of crops to water stress. Small changes in the water level of Lake Chad can cause significant variations in the amount of land which is flooded each year. A delicate balance is often struck between the two environmental extremes of drought and flooding. Farmers have learnt through experience that crops with longer gestation periods and which are adversely affected by inadequate soil moisture during the critical seeding period will not survive. Farmers grow both drought-tolerant *masakwa* sorghum during the dry season and flood-tolerant crops such as maize and cow-peas during the wet season. They also experiment with different *masakwa* varieties, each with its varying vulnerability to risk. If there is excessive flooding, the flood tolerant varieties will survive, and in the event of a lack of rainfall, *masakwa* will produce yields. As a risk-aversion strategy, the combination of flood-retreat and flood-advance cultivation techniques practised by the Kanuri has been remarkably successful (see the table below):

Farming techniques developed by the Kanuri

Zuru (flooded zones)	Kudu (non-flooded zones)
Kati Kassa sandy soil cultivation	*Kati Kassa* sandy soil cultivation
Kati Kulum silt and clay cultivation	*Firki* clay cultivation (around Lake Chad)

However, *masakwa* cultivation has not always been so prevalent. Successive state land policies promoting rural development through state interventions have had a serious impact on *masakwa* cultivation in Borno state.

HISTORY

During the last decade of the colonial period (1950s), *masakwa* cultivation took place on an unprecedented scale, partly due to the favourable climatic conditions, and partly to the relatively high lake level which resulted in considerable flooding of the area. Irrigation activities were limited to the control and regulation of floodwater (Hailey, 1957), and the development of irrigation schemes was approached cautiously. The technical feasibility and economic efficiency of large-scale irrigation in Nigeria was thought to be dubious in the light of the poor performance of various irrigation projects elsewhere, such as the enormous Gezira cotton-growing scheme in the Sudan.

From 1973 to 1984, there was a massive abandonment of *firki* soil cultivation. This was partly due to years of less-than-average rainfall but, more importantly, large-scale irrigation projects were developed. Most notable among these was the South Chad Irrigation Pilot Project which was established in 1970 to provide help in the planning and design of a bigger project, the South Chad Irrigation Project (SCIP). The aim of the latter scheme was to irrigate 67,000ha of land in order to reduce the ecological deterioration of the region and the risk of drought, with the anticipated benefits of increased stability and scope for increase in family income and food supplies at both the local and national level.

By 1973, the government had completed the plan to acquire about 100,000ha of *firki* soil on which *masakwa* is traditionally cultivated south-west of Lake Chad. Around 55,000 farming families were resettled as a result of the scheme and *masakwa* cultivation was largely abandoned by the Kanuri. Thus, rather than improving the production efficiency of the indigenous agricultural techniques, the government imposed an irrigation scheme that was very large, sophisticated, capital intensive and geared towards the production of import-substitution crops such as rice and wheat.

State intervention also profoundly altered the traditional land tenure system described earlier. The Land Use Act of 1978 transferred land to the Chad Basin Development Authority (CBDA), a government parastatal and the parent body of the SCIP.

The fall and rise of *masakwa* cultivation

When the SCIP was eventually commissioned in 1979, only 500ha of land were irrigated, and although by 1983 10,000ha were irrigated, there has been no increase since then. Why has this irrigation scheme been so unsuccessful? Rainfall in the Sudano–Sahelian zone has fallen below the 1930–60 mean, with lows in both 1972–73 and 1984–85 (Grove, 1985). The period 1983–85 was the driest this century, with the lake falling to its lowest level and shrinking to its smallest size, from a surface area of 23,000km^2 in 1962–63 to just 3,000km^2 in 1984.

Environmental changes of this scale have imposed overriding constraints on the operation and management of all irrigation projects dependent on Lake Chad. Water supply has been highly erratic and inadequate. In 1981, massive crop losses were reported on all the Nigerian irrigation projects in this area. Consequently, the majority of the farmers had to abandon their fields because of the increasing uncertainty over water supply and the low rate of return. By 1983–84, all irrigation schemes in this area had been adversely affected by water shortage.

The near-collapse of the SCIP has led to a resurgence of indigenous soil and water conservation practices, and *masakwa* cultivation on the *firki* clay soils around Lake Chad has resumed. This has been accompanied by the movement of farmers on to the lake floor to produce staple food using the residual soil moisture of the lake.

Masakwa cultivation in practice

Masakwa cultivation is labour intensive owing to the need for earth bunds. Water accumulates and infiltrates, raising the level of the water-table and recharging the aquifers. Made with heavy clay soils, they are impervious and erode slowly, so any rainfall is trapped and retained in the soil.

The construction of earth bunds, or strips, and pits requires a heavy investment of labour in both their construction and maintenance. Men are mostly involved in their construction which is usually carried out by individual land-users using family labour, depending on the scale of the operation and the scope of the farming operation. Kanuri women also farm and are actively involved in other phases of *masakwa* cultivation, such as sowing, weeding, harvesting, processing and marketing. Women with *masakwa* fields beside their husband's fields also use soil and water conservation techniques on their fields. However, household labour is not only insecure (because of the high rates of divorce), but also extremely unpredictable. Kanuri farmers therefore have had to look to alternative sources of labour supply, either by using clients or labourers, or by taking part in co-operative activities. During a good rainfall year, it is possible to cultivate the full extent of the bunded area, and water is retained in the unbunded, communally owned plains and depressions to allow them to be cultivated as well.

A non-waterlogged area is usually selected as a nursery site. Seed-beds are tilled with hand-hoes, and weeds are collected and removed from the site. During the land preparation and immediately after the flood recession in September, the grasses in the field are slashed to the ground and allowed to dry. Any excess grasses are then spread on the soil surface and burnt. This, together with green manure supplied by mulching, is the main source of organic fertilizer which is applied to the inherently fertile vertisols.

The seeds are spread evenly on the seed-beds in mid-August and the seedlings are allowed to grow for six weeks before they are transplanted to the main field during September. Healthy seedlings are carefully uprooted from the nursery and their leaf tips are snipped off before they are transplanted to reduce transpiration. Seedlings are planted in the pits and about 200ml of water is poured into each pit before two seedlings are inserted. The pits are left uncovered and no fertilizer is used. During the rains, the clay soil cracks and absorbs the water. The *masakwa* growers conserve the moisture by pulverising the top soil with hoes during weeding to seal up the large cracks created when the clays expand and contract. Thus, most of the water that might otherwise be lost through evaporation is conserved.

A low plant density is adopted by the farmers to avoid diseases such as covered smut (*Spacelotheca sorghi*), loose smut (*S. cruenta*) and heat smut (*S. reiliana*) (Norman *et al*, 1984). Narrow rows require herbicides for effective weed control, and weed control appears to be very important in *masakwa* farming (Myers and Foale, 1981).

Some farmers leave their crop residues after the harvest, while others deliberately cover their plots with grass brought from the unfarmed areas which improves the soil structure and also reduces water run-off, and allows more water to infiltrate into the soil. Harvesting is done manually from mid-January when plants are cut at ground level and laid flat to dry out in the field before the panicles are removed and then threshed to obtain the clean grains. Yields are generally low and variable, with an average of 500–800kg/ha, good rainfall bringing a larger harvest.

The traditional farming systems contribute significantly to self-sufficiency in food production. Although the average yield of *masakwa* per ha is quite low, nevertheless it complements production from other sources. Farmers thus regard *masakwa* cultivation as their lifeline, most importantly because it allows farmers to grow crops at a time when practically no other crop could be grown in this area without irrigation.

As a survival strategy, farmers follow a series of activities according to the season. From February to June each year, for example, farmers may engage in market gardening as well as participating in large-scale irrigation projects. From June to September they may engage in upland rainfed farming; from August to January, the traditional dry season, drought- and cold-tolerant sorghum, *masakwa*, is grown on the *firki* clay vertisols surrounding Lake Chad, and from February farmers move on to the lake floor to farm again in the flooded zone. For some farmers, lake-floor farming is becoming a year-round activity as multiple cropping is made possible, depending on the rise and fall of Lake Chad each year (Kolawole, 1988).

But can this practice be sustained? The continued cropping of *masakwa* on *firki* soils will depend on a number of environmental, economic and social factors. The most important constraint to successful crop production in this environment is water, which places more severe constraints on yield than the availability of crop nutrients (Norman *et al*, 1984). Traditional farming systems in this zone have always been determined by the seasonal variations in the level of Lake Chad, which is itself determined by rainfall over the surface of the lake, the discharge of the lake's principal affluents and the rainfall conditions in the catchment areas from which they both take their sources.

Insects, pests and plant diseases such as the migratory locust, the desert locust, grasshoppers, army- and headworms can cause crop losses as high as 25–45 per cent (Davies, 1985). There are also various plant diseases, such as stem and leaf rusts, foot and root rot, blight and blast (*Piricularia*) in wheat, while damage to crops from birds can be a major problem.

Transport to the market is problematic as few vehicles pass through the area, increasing the cost to farmers of bringing their produce to the market and making the transport of highly perishable goods such as tomatoes, fruit and vegetables even more difficult.

The youthful population and high percentage of women, coupled with a high out-migration rate for men also has important implications for agricultural development. There is a clear labour shortage for agricultural activities and limits on capital investment for agricultural production. Little effort has been made to improve the technical efficiency of *masakwa* cultivation or to intensify agricultural practices in the way of improved fertility management. Whatever efforts are made are confined largely to the mulching and zero tillage practices which were already in use in the area. No chemical fertilizer is used nor organic manure applied.

Lastly, although *masakwa* plays an important economic role in the major producing areas, out of an estimated 412,000ha of vertisols in Borno, only about 10 per cent are presently cropped to *masakwa*. The farming capacity of the Kanuri could be increased considerably by expanding the production of *masakwa* cultivation across the state.

CONCLUSION

Farmers have thus learnt to combine *masakwa* cultivation, lake-floor farming, small pump irrigation and rainfed upland farming with irrigated farming, whenever available, for both home consumption and the market. *Masakwa* cultivation and indigenous methods of soil and water conservation will continue to play a vital part in the successful farming practices of the Kanuri in this area, as long as state interventions such as the South Chad Irrigation Project continue to fail.

12

INDIGENOUS SOIL AND WATER CONSERVATION IN SOUTHERN ZIMBABWE

A study of techniques, historical changes and recent developments under participatory research and extension

J Hagmann and K Murwira

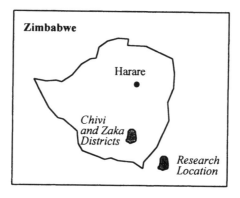

Soil and water conservation (SWC) have played an important role in the development of smallholder agriculture in Zimbabwe. The introduction of the plough early this century triggered a drastic change in the farming system and in practices, virtually eradicating indigenous technical knowledge in soil and water conservation. As a consequence, degradation became so severe that from the 1940s the colonial administration forced smallholders to implement externally developed conservation works.

At present, there is growing concern about continuing degradation and alternative SWC systems are being studied (Norton, 1987; Elwell, 1993). This paper analyses indigenous soil and water conservation with a view to promoting appropriate SWC techniques in smallholder agriculture in semi-arid Zimbabwe. The study was carried out in Chivi and Zaka districts in southern Zimbabwe. The area is marginal for cropping with poor sandy soils and erratic rainfall. Soil and water conservation activities have been carried out by two projects since 1991: The Intermediate Technology Development Group (ITDG, a UK-based NGO) Food Security Project with its main emphasis on the extension of SWC techniques and the Agritex/German Technical Development Agency (GTZ) Conservation Tillage Project which is more research orientated.

For an assessment of techniques and the historical development of the farming systems, semi-structured interviews guided by a questionnaire were carried out, covering changes to the traditional farming system. Results of these surveys were discussed in workshops, with farmers paying particular attention to the role of local institutions with regard to SWC.

A HISTORY OF INTERVENTION AND CHANGE

Agriculture before the introduction of the plough (c 1920) was based on livestock and shifting cultivation. Livestock provided food, clothing, transport and manure but not draught power, and had a major role in the social system.

Finger millet was the main crop, together with sorghum, pearl millet, groundnuts and other minor crops. Land was cleared from the bush with hand axes and cultivated for three to ten years. A fallow period which allowed soil fertility to be restored followed and a new piece of land was cleared after shifting. Wetlands were also cultivated and provided a major source of food during drought years. The collection of wild fruits and the hunting of wild animals contributed considerably to food security.

Traditional rules ensured the sustainable use of natural resources. The burning of trees without a good reason was prohibited. Water sources were mostly sacred and therefore protected from pollution. The permission to cultivate a new piece of land came from the traditional authorities, who enforced these rules together with the spirit mediums.

With low population density and limited tools, traditional soil management prevented severe soil degradation. For example, when bushland was cleared with hand axes and hoes, tree stumps were left in the field and ash spread on the field, while soil disturbance was minimised through shallow hoe cultivation. A considerable area of the drylands was planted on hand-made ridges. Intercropping developed a dense soil cover so that erosion hazard after crop establishment was minimal, and the high percentage of ground cover reduced soil evaporation and therefore drought vulnerability. Further measures included mulching with weeds or burnt crop residues, construction of soil bunds, stone bunds and ridges, and management of wetlands for rice production in the wet season and maize on big ridges in the dry season.

Changing towards the plough and maize with the emerging extension service

The introduction of the plough and maize as a food crop came with the arrival of white settlers at the end of the last century and became adopted on a large scale by indigenous farmers between 1920 and 1940. Improved weed control allowed an expansion of the cultivated area and attractive producer prices for maize were an incentive for the expansion of maize production, triggering the market production of maize.

With the introduction of the plough and maize, agricultural extension started in the 1930s. A package of cropping practices was developed by an American missionary (Alvord, 1958; Page and Page, 1991) and is still promoted today. This included the utilization of manure, crop rotation (cereal crops with legumes), row planting and mono-cropping, autumn and pre-plant ploughing, as well as the uprooting of trees in the field to provide for easy tillage and straight planting lines. Alvord was convinced that the best way to improve farming and conservation was to consolidate arable holdings, to separate these from grazing lands by perimeter fencing and to locate villages along the margins of the cultivated areas.

The expansion of the area under cultivation and increasing population pressure resulted in longer cultivation periods. The shifting of fields was reduced and later abandoned completely. The recovery of fields during the bush fallow period took longer after the complete clearance of woody vegetation.

Two-thirds of the respondents noted an increase in soil erosion after the introduction of the plough. Rill and gully formation increased particularly. However, all still consider the system positive as it increased work rates and efficiency.

Imposition of externally developed contour ridges to fight accelerated erosion

By 1930 erosion and degradation had become so severe that the then colonial government promoted conservation works and enforced by law highly unpopular measures, such as the digging of contour ridges and destocking (Native Land Husbandry Act, 1951). It also prohibited stream bank cultivation (mostly gardens in wetlands), which was a major threat to food security in drought years.

Contour ridges are part of a conservation layout designed for commercial farming areas. These are normally sited in high rainfall areas and on heavier soils, aimed at stopping erosion by draining excess water off the fields. Despite different climatic conditions, these structures were imposed in semi-arid areas as well, where water retention would have been appropriate.

The introduction of contour ridges meant the end of shifting cultivation and bush fallow. Permanent cultivation under the plough and the adoption of the Alvord extension package increased considerably, partly as a result of the incentives promised. The package was also partly adopted as a result of land use intensification necessitated by the increasing population. The technical design of contour drains which drained excess water led to many farmers rejecting the measure. Farmers were punished in an attempt to make them willing to implement the measures, resulting in poorly constructed ridges, which can increase gully erosion.

Relaxation of the enforcement of SWC and commercialization of smallholder agriculture

During the liberation struggle (1976–80), farmers were encouraged by the freedom fighters to destroy or stop maintaining contour ridges as a symbol of white oppression. After independence in 1980, it became very difficult for the new government to enforce the conservation objectives. The major focus of agricultural policy and extension became the commercialization of small-holder agriculture using high-input technology (hybrid seeds, inorganic fertilizers and chemicals for pest control) which was previously available only to the large-scale commercial farms. The focus on conservation in extension lost out to the drive towards yield maximization, and the maintenance of conservation works was neglected (Whitlow, 1988). The traditional soil and water conservation methods such as intercropping were discouraged. The productivity decline due to severe erosion was buffered by the application of manure and inorganic nutrients. However, several severe droughts between 1980 and 1993 have reduced the number of cattle to a minimum, implying that now negligible amounts of manure are available and it is doubtful whether it will be possible to sustain the present levels of production with inorganic fertilizer.

The changes in the farming system have almost eradicated traditional conservation measures. Conservation is understood by farmers and extension workers to be a synonym for contour ridges (Hagmann, 1996).

In workshops, farmers' perception of degradation and the need for conservation was analysed with regard to the change in the state of their natural resources during the last 20 years. The results were identical in both areas. There is a common perception that the natural resource base has dwindled drastically over the last two decades and that the situation will continue worsening unless stringent measures of resource management are put in place. But traditional leaders who used to enforce laws on SWC have been stripped of their powers in favour of rather weak village development committees (VIDCOs). All the respondents now emphasize the importance of soil and water conservation, based on the fact that 'there is no place where we are going to move to after the final degradation of our land', or 'we wish we could be taught ways of maintaining those small fields we are left with'. Most people realize the need for resource management, including soil and water conservation, but feel helpless to organize themselves properly to define and implement appropriate conservation measures.

The role of indigenous SWC in the present environment

The plough and Alvord's package for permanent agriculture (Alvord, 1958) provided a technology which, at least on a short- and medium-term basis, addressed the requirements well and was therefore very successful. The major mistake, however, was the contempt for traditional knowledge in favour of modern Western technology. The new technology was not based on traditional, sustainable soil management systems which could have been improved by integrating new components, but on the superiority of a completely new system imported from the temperate climates of Europe and North America. In a relatively short time, this approach managed to wipe out indigenous soil and water conservation systems and farmers' confidence in indigenous agriculture in general (Page and Page, 1991). Today, three generations later, the plough is considered 'traditional' practice by most people in Zimbabwe and the Shona word for cultivation even means 'ploughing'.

Techniques used in the last century are largely inappropriate to the present population pressures on land. Nevertheless, the principles which were effective in the traditional system, such as soil cover, minimum tillage, planting on ridges, intercropping, soil bunds and stone bunds, would be taken as the basis for developing improved techniques. A synthesis of traditional techniques and new methods for SWC could be adapted to specific sites, situations and farmer needs.

The role of local institutions and government authorities in SWC

The major changes in the farming system described above were accompanied by a drastic change in social organization. Before the establishment of the colonial system at the end of the 19th century, few local institutions existed. The major institutions at that time were spirit mediums (considered as messengers of God), traditional healers and the chiefs, the last of whom were responsible for land allocation and management.

Removal of power from the traditional leaders after independence made them feel that they were no longer responsible for soil and water conservation, and thus many of the conservation laws were diluted and no longer executed. The latest trend, after chiefs and traditional leaders have regained power for land allocation, is a monetarization of traditional rights and laws which is corrupting the traditional leadership. Cases in which kraal heads sell land which is highly vulnerable to soil erosion to immigrants are on the increase. The result is increasing land degradation, the opposite of the objectives they pursued previously.

Present functions and the capacities of local institutions and of government authorities in SWC

The importance of different institutions varies in each location. Traditional leaders (kraal heads and chiefs) are considered most important in the management of natural resources and SWC. The involvement of VIDCOs, Village Community Workers (VCWs) and agricultural extension workers ranks far behind them.

Farmers' expectations of their institutions concern their role in preventing the burning of vegetation, the cutting of trees and the cultivation of stream banks. Traditional laws on resource management are perceived as rather contradictory: a third of those interviewed still hold by them, while others said that many of these laws no longer exist. The position of the chiefs today is much weaker, with 40 per cent of respondents pointing out that they do not take any steps if traditional laws are broken, while 60 per cent claimed that although they are still active, their power has been overshadowed by modern institutions like VIDCOs.

Implications of the local institutional set-up for interventions in ISWC
The institutional set-up at local level is highly complex, the social and generational conflicts being reflected in the generally weak leadership and co-operation found within rural society. Traditional leaders and structures are at odds with those who are in favour of modern democratic institutions. In such a complex and tense situation, the major condition for success with indigenous soil and water conservation (ISWC) is to work with local institutions and to strengthen them.

RECENT DEVELOPMENT UNDER PARTICIPATORY RESEARCH AND EXTENSION

Participatory approaches to SWC

The ITDG Food Security Project and the Agritex/GTZ Conservation Tillage Project have both been following participatory approaches to their work from 1991 onwards, with an emphasis on the strengthening of local institutions as vehicles for any developmental activities.

Using a community awareness raising approach, 'Training for Transformation', TFT, the importance of participation and co-operation are emphasized in organizational development in order to build institutions which enable people to become self-reliant. It manages to integrate and unite often conflicting interests under one umbrella. As described earlier, change within Shona society has weakened the social coherence and security which was based on traditional rules and regulations. A new 'umbrella' could provide a means to resolve and promote innovation. This is particularly important for SWC where all land users in a watershed must agree to implement certain measures if they are to succeed.

In both projects, several options for SWC have been developed by farmers. Others were developed on station and were offered to farmers for testing. Most of the options have their origin in traditional farming practices, but are adapted to the present farming system (see photos 16 and 17).

Mechanical conservation options

- Stone bunds as check-dams in rills and small gullies along the contour line.
- Infiltration pits in the contour drain to retain water and soil from flowing out.
- *Fanya-juu* terraces for maximum water and soil retention in the field.

Agronomic conservation

- Weeding systems to enable a reduction in ploughing.
- Conservation tillage: tied ridging and mulch ripping.
- Intercropping.

Biological conservation

- Compost.
- Vetiver grass for rill reclamation and grass strips.

Water-saving irrigation methods for gardens

- Sub surface irrigation with homemade clay pipes (Murata *et al*, 1993).
- Underground plastic sheets to prevent deep percolation.
- Inverted bottles directed to roots to reduce water loss due to evaporation.

Other less widespread options which are being tested include agroforestry and wetland management. As problems and needs differ from area to area, not all the options are being promoted for testing in all areas.

Impact of the participatory approach

The impact of the approaches taken in the two projects has been evaluated in terms of the technological achievements and the effects on local institutions and social organization. In both cases, the results are very encouraging and should be further monitored during the coming years.

Farmers were asked about their perception of the old extension approach, compared with the new participatory approach to developing soil and water conservation techniques. Three major differences need to be highlighted. First, farmers now feel that everybody can participate in soil and water conservation, rather than being limited to Master Farmers. The second major difference is the process of dialogue which is now encouraged, involving the explanation of soil processes rather than the imposition of solutions on farmers. Farmers also noted that they were now encouraged to co-operate and to share knowledge between themselves and the researchers (see the table on next page).

The participatory process seems to have initiated a major drive towards improving the leadership of local institutions. With leadership and co-operation being one of the major problems often mentioned by farmers, these results indicate that institutional strengthening has had a positive effect on farmers' capacity to organize themselves and to increase participation in agricultural development through club membership. In one ward, membership of farmers'

Farmers' perception of the old and the new approach and the most important aspects of the participatory approach

Characteristics of approaches

Old approach
Forceful methods were used.
Only few people could benefit (eg, literate).
Intercropping was forbidden.
Failed to address SWC convincingly.
We were told to do things without questioning.
Usefulness of conservation works never explained.
No dialogue between farmers and extensionists.
Little co-operation among farmers.
Extension agents treated our fields as theirs.

New/participatory approach
Everyone to benefit as all are free to attend meetings now.
There is dialogue.
Process is well explained (teaching by example).
Farmers are the drivers now.
Intercropping is encouraged to boost yields.
Farmers are being treated as partners and equals.
No discrimination against poor or rich, educated or uneducated.
We are given a choice of options.
They pay attention to us and take time to find solutions to farmers' problems.
We are being encouraged to try out new things.
It helps farmers to work co-operatively.
Farmers practise SWC with enough knowledge of why they should do it.
Learning from others through exchange visits/learning through sharing.
It helped farmers to develop the ability to encourage each other in farm activities.
Encouragement to practise SWC through various options.
It is capable of mobilizing large numbers of people.
The approach brings about desirable SWC techniques through participation.
Farmers are free to ask for advice.
Yields have increased through SWC techniques.
The dedication of modern extension agents/researchers.
It has brought development in the area.
It is very effective in the conservation of trees, soil, water.

clubs, which are the organizations that carry out SWC, has increased from 120 in 1991 to more than 800 in 1994. This was mainly due to changes in the club leadership as a result of the process of the strengthening of local organizations. The villagers managed to choose and elect leaders who were popular and had the potential to motivate people.

Impact on development and the spreading of SWC techniques

Within Chivi, with a total of 1136 households, at least 80 per cent are practising

SWC in one form or another. The range of technologies currently in practice include mulching, tied ridges, the use of clay pipes, plastic sheets and inverted bottles for the irrigation of gardens, infiltration pits, intercropping and rock-catchment water harvesting. The adoption of the different techniques during the first three cropping seasons is shown in the table below:

The adoption of SWC techniques in Chivi (Ward 21)

Technique	Adopted by number of farmers			Source of technique
	1992–93	1993–94	1994–95	
Cropped fields				
Tied ridges/furrows	28	>100	>500	ConTill Project/ Chiredzi Research Station
Infiltration pits	20	289	>800	Farmer innovation (Mr Phiri)
Fanya-juu bunds	0	4	nd	ConTill Project
Mulching	2	3	nd	ConTill Project
Intercropping	~50	>450	nd	Indigenous farmer knowledge
Spreading of termitaria	78	>128	nd	Indigenous farmer knowledge
Tillage implements	0	96	nd	ConTill Project
Gardens				
Sub-surface irrigation/ garden	~50	68	nd	Chiredzi Research Station
Plastics/inverted bottles	10	>200	nd	Unknown farmer
Compost in gardens	4	14*	nd	Farmer/ConTill Project
Mulching in gardens	85	>300	nd	Farmer

*Groups out of a total of 37 practising this technique.

Because of very limited animal draught power, tied ridges continue to be constructed by farmers by hand through labour parties. Infiltration pits have become very popular and have spread beyond the ward through farmer to farmer sharing, whereas *fanya-juu* bunds have only recently been introduced. Mulching is particularly popular in gardens where 60 per cent of the group members are practising it for water-saving purposes. Spreading of anthill material is a traditional technique which has been revived. Testing of tillage implements only started in the 1993–94 season. The focus is on draught power-saving implements and the use of donkey power. In general, SWC techniques for gardens are most popular as they contribute directly to a reduction of labour and increased production through the prolongation of water availability in the wells. As gardening is a major source of income generation in the area, these techniques are particularly important.

In the ConTill Project experimentation among farmers has spread to include other subjects linked to SWC, creating numerous ideas and farmer trials. Trials on topics such as mechanical conservation works, rill reclamation, live hedges, fodder plots, new varieties, planting techniques, plant spacings, intercropping, strip cropping, mulching, wetland cultivation and composting have been thought out in detail by farmers and are being carried out.

An interactive process between researchers and farmers has been established which has led to increased self-confidence and the building up of knowledge based on experience. This is manifest in farmer-organized field days to share knowledge with their neighbours. Researchers and extensionists are invited as guests, not as experts.

CONCLUSION

The introduction of the plough more than 70 years ago brought about major changes to the agricultural system and the abandonment of many ISWC techniques. An extension policy which assumes the superiority of Western technology over African agriculture has reduced farmers' confidence in their own solutions. Social organization has been weakened as traditional and modern institutions have been juxtaposed, while conflicts between generations have worsened as a result of socio-cultural changes.

Participatory processes have been used to revive and combine indigenous knowledge and research capacities of the local farming communities with that of research and development institutions in an interactive way. By working with and strengthening local institutions, farmers' confidence in their own capacity for experimentation has created a new generation of soil and water conservation techniques which builds on traditional knowledge but is adapted to current conditions.

13

ENVIRONMENTAL CHANGE AND LIVELIHOOD RESPONSES

Shifting agricultural practices in the lakes depression zone of Northern Zambia

Patrick M Sikana and Timothy N Mwambazi

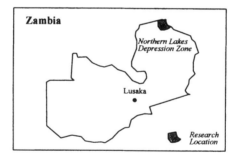

When this research began, the first instinct was to go out in the field to look for elaborate physical structures that had been constructed and managed by local people. Kaputa district, which has a dry hot climate and also happens to be one of the least developed areas in the country, was chosen. It was assumed that the harsh environment would provide fertile ground for indigenous skills and innovations to flourish. At the end of the first day in the field, the atmosphere back at the camp was one of gloom and despair for no bunds, terraces, contours or any such physical structures for harvesting water or controlling erosion had been discovered, and yet it was known that moisture stress caused by insufficient and erratic rains is one of the most important limiting factors to agricultural production in the district.

Resisting the temptation to blame the local people for a lack of imagination, it was resolved the following day to be a little more imaginative. It was realized that it was a mistake to assume that technology exists more in terms of physical structures than to look at indigenous soil and water management regimes in terms of 'performance' (Richards, 1989). It is important to look at a 'range of strategies' rather than a single technological choice, and to acknowledge the importance of social differentiation in determining what strategies are suitable for different people.

In this paper, it is argued that the current debate on indigenous soil and water conservation (SWC) has not paid due attention to the fundamental issue of 'survival'. Eagerness to promote 'proven' indigenous SWC techniques has often meant that the special character of local people's needs has been ignored and their capacity to deal with ecological change has been underestimated. The tendency to 'scientise' (Thrupp, 1987) and package indigenous SWC for widescale dissemination may not work because the success of indigenous SWC does not depend solely on the scientific efficacy of the technology itself. It must also be relevant to the socio-economic and historical situation of the local community in which the technology is applied. In this context, local participation is desirable, not merely for ideological reasons of 'empowerment' but to ensure that SWC strategies are in line with local priorities and are sustainable.

INTRODUCTION TO THE STUDY AREA AND PEOPLE

Kaputa district is situated in Northern province, the largest of Zambia's nine provinces. Owing to its relatively low elevation, the climate is subtropical. The study area includes the whole of Kaputa district and the north-western parts of Mbala district within the northern lakes depression. This zone is one of the few areas in the Northern province which has a population density of over ten people per km^2 of suitable agricultural land (ARPT, Northern Province, 1986). According to the 1980 census, the population of the district was 44,344 people, out of whom 97 per cent were rural dwellers. By 1990 the population had swelled by 45 per cent to around 66,000, far outstripping the rate of growth of neighbouring districts.

The influx of people into Kaputa district which started in the 1950s stemmed largely from the expansion of the fishing industry. With the declining availability of fish in neighbouring Lake Mweru of Luapula province, people were forced to look for new fishing grounds along the western shores of Lake Tanganyika and around Kapinda Lagoon in Kaputa. According to the local people, the present lake around which this study is based did not exist at that time. In its place was only a small stream which fed into the Kapinda Lagoon. Successive heavy rains in the early 1960s led to a gradual expansion of the stream and the Kapinda Lagoon into a big lake which came to be known as *Mweru na ntipa* (Mweru with mud).

Previously, fishing was carried out on a very small scale for subsistence purposes. However, the expansion of the lake and the introduction of new fish species led to a further influx of people from neighbouring districts and provinces and from as far afield as Zaire, Tanzania and Malawi. Coupled with dwindling employment opportunities in the major towns along the railway line, the number of young people leaving the area to look for wage employment has fallen. As a result, the population of the lakes depression zone has grown rapidly. This has led to strategic changes in patterns of resource utilization influenced both by climatic conditions and shifting vegetation. Participatory Rural Appraisal (PRA) exercises were conducted to obtain information about short- and long-term rainfall trends in the area. The years 1992 and 1993 were particularly dry. This led to a drastic decline in rice cultivation. Going further back, it emerged that rainfall had increased during the late 1950s, reaching a peak in the early 1960s. This led to the expansion of the lake. During this period, drought was not so severe and relatively abundant amounts of finger millet could be grown. However, the impact of drought worsened during the 1970s as rainfall decreased. Since then, rainfall has either remained constant or continued to decline, never returning to pre-1960 levels.

The vegetation in the district changes with declining altitude and the drier climatic conditions. The *mutengo* woodland at the higher levels, which is interspersed with scattered shrubs, gives way to *mateshi*, a thicket-type vegetation. This extends towards the lower slopes which are characterized by *chipya* or what is described by Trapnell (1953) as 'high grass woodland'. This is the result of the destruction by fire of earlier dense forests (Mansfield *et al*, 1976). Combined with these major vegetative niches are a number of recurrent topographical features around the lake such as marshes, seasonal mud-flats and *dambos* called *ilungu*. Each of these micro-environments poses different problems and opportunities for farming.

Due to human activity, the vegetation around the lake has been shifting boundaries on a continuous basis. In response to land cultivation and lake expansion or contraction, these vegetation types fluctuate in size. According to the local people, around the 1960s the lake expanded and submerged some of the *ilungu*, thereby substantially reducing its size. More recently, the *ilungu* has expanded again as a result of the erratic rains and the consequent contraction of the lake. With cutting, burning and cultivation, the *mateshi* has dwindled in size allowing the *chipya* to spread. The *mutengo* woodlands are often protected areas and so tend to remain relatively constant in size. Using a PRA exercise, a schematic illustration of these dynamics as a function of time was obtained (see the figure on next page).

The study area consists predominantly of sandy soils with a low water-holding capacity. *Chipya* soils have traditionally been considered by local farmers to be less then fertile and of little use except for growing cassava, a crop that is able to thrive under marginal soil conditions. However, although relatively poor, the value of *chipya* soils is considered high today because of the heavy reliance on it for food production. As a result, local SWC strategies to exploit *chipya* soils have spontaneously developed. The table on page 111 describes the major characteristics of the vegetative areas surrounding the lake.

The changes caused by the interaction between humans and the natural environment are characterized by a continuous process of adaptation. Human

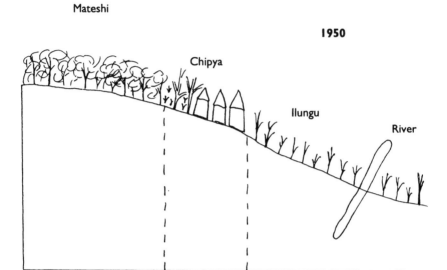

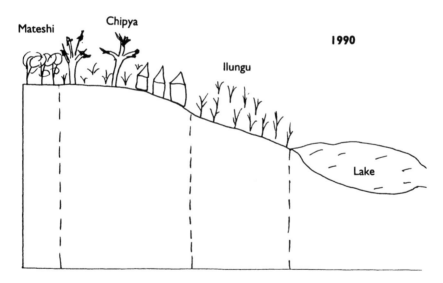

Participatory mapping by villagers of shifting boundaries of the major land systems in the lakes depression

Characteristics of vegetative areas in the lakes depression zone

	Ilungu	Chipya	Mateshi	Mutengo
Soils	Covered with water, depending on rains	Sandy, less fertile, poor water-holding capacity	Rich, clayey, retains moisture well	Heavy, rich soil
Vegetation	Marshes, mud-flats, *dambos*	Grasslands	Scrub, evergreen thickets	Forest
Use/ management	Rice (in wetter years); maize, modern vegetables, sugarcane (in drier years)	Cassava; maize, as farmers evolve new moisture conservation strategies (mounds)	Finger millet, with maize and cassava as ancillary crops (with or after finger millet)	Not cultivated, used for hunting

activity causes change in the natural environment and the resultant change in turn requires adaptation in human activities, and so on. The fragile and difficult environment of the lakes depression zone is increasingly required to support more and more people. When environmental resources are subjected to increasing pressure, new and more efficient forms of resource use will spontaneously develop. This is examined below.

FASHIONING NEW LIVELIHOODS IN A CHANGING ENVIRONMENT

The last section painted a picture of environmental disaster. The *mateshi* thickets that were used previously for finger millet cultivation are being depleted, rich soils are gradually being converted into less fertile *chipya* soils, and the rains, in the eyes of the local people, are becoming less predictable. The land, meanwhile, is increasingly required to support more and more people. The range of strategies used by different categories of people to deal with these ecological challenges is examined.

Since moisture stress is an important variable in the production system, many of the adaptation strategies are geared towards more efficient utilization of the available moisture. The choice of a given moisture utilization strategy depends very much on the socio-economic circumstances of different categories of people, and will often be influenced by the opportunities and constraints presented by alternative sources of livelihood.

Some 50 years ago, finger millet cultivation based on the *mateshi* slash-and-burn system was the predominant system of land use in the lakes depression zone (Trapnell, 1953). Finger millet grown in this manner could sustain the small population, although famine resulting from drought-induced crop failure was also very common. Shortfalls in food were made up by importing finger millet from the neighbouring upland district of Mporokoso. Barter was the main form of exchange, with products from Kaputa, such as fish, salt, wild game meat, baskets, etc, exchanged for grains. However, as the population increased in the 1950s, this strategy was no longer feasible. During poor harvests demand for grains far outstripped supply. The community thus had no choice but to devise new coping strategies.

Greater demand could not be met by intensifying finger millet production for three reasons. First, increased population put immense pressure on the traditional slash-and-burn system, leading to the progressive depletion of the *mateshi* and its transformation into less fertile *chipya*. Secondly, owing to its small grains, the caloric value of finger millet per unit of land was insufficient to meet increased needs. This problem was compounded by an inability to increase the area under cultivation. Thirdly, the *mateshi* slash-and-burn system of land use had to compete increasingly with fishing for household labour as fishing became more commercialized. This engaged a larger proportion of the productive male labour in the community.

This called for more efficient strategies of land use. The new strategies of adaptation were those which could respond to the challenges of erratic rains and the need to utilize the less fertile but more abundant *chipya* soils, the need to balance labour demands for the dual enterprises of fishing and farming and, above all, the need to provide sufficient caloric intake for greater numbers of people.

New crops: cassava

Cassava is now the most important staple in Kaputa district, with almost every household involved in cassava cultivation (ARPT, Northern Province, 1986, 1987). Cassava was grown in the area as far back as the 1930s. However, the indigenous varieties grown were late maturing, very bitter and only grown on a very small scale, usually intercropped with finger millet in *mateshi* gardens. Since these varieties were very bitter, they were only used for supplementing finger millet during lean times.

Around the mid-1940s, new varieties were introduced into the area by Zairean immigrants. By the 1980s, cassava was the major starch staple grown in the area and was no longer confined to *mateshi* fields. Most of the cassava is now grown in *chipya* fields as only a small proportion of the population is able to continue with *mateshi* cultivation. The shift of crop types from finger millet to cassava and land use from *mateshi* to *chipya* presents both new opportunities and challenges.

Cassava provides answers to endemic drought risk, increasing human population, competition between fishing and farming, and dwindling *mateshi*. First, it is well adapted to prevailing dry conditions and can therefore be cultivated on a relatively large scale. Secondly, cassava performs reasonably well in the heavily depleted *chipya* soils which cannot support finger millet without costly inputs such as fertilizer. The high-yield potential of cassava on marginal soils allows it to feed more people than finger millet or other staples. Thirdly, cassava growing is less labour demanding than the traditional *mateshi* slash-and-burn system. The move from slash-and-burn farming to cassava gardens reduced by as much as 40 per cent the labour needed to satisfy annual food requirements (Holden, 1993). This is of great advantage in the lakes depression zone where the brunt of agricultural labour is borne by women and older men.

The other reason for the increased importance of cassava is the concentration of population in the major fishing centres and the increased commercialization of the fishing industry. The demand for cassava among professional fishermen and fish traders has created a market for cassava, once a purely subsistence crop. Whereas both fishing and non-fishing households grow cassava in all villages of the study area, there are large fishing centres where professional fishermen working for big fishing companies do not cultivate crops at all.

Trading cassava enables some households and some categories of individuals, such as women who do not fish, to participate in the cash economy. Although women's access to cassava has to be negotiated, nevertheless they exercise a great degree of control over cassava consumption and disposal because all of the post-harvest tasks, such as uprooting, peeling, soaking and drying, are undertaken by them. Trading in cassava is also an important strategy for younger men who aspire to earn income for investing in the more lucrative fishing enterprise. Since cassava requires minimal attention during most of its growing period, younger men can combine cassava growing with trading and fishing, without difficulty.

Maize production

With the decline of finger millet, maize has become an important cereal staple which complements cassava. Maize is not new in the study area. In the past, maize (mainly of the local open-pollinated flint type known as *kalinwa*) was grown on a very small scale, either sparsely scattered in the *mateshi* finger millet gardens or cultivated on richer and wetter *ilungu* soils. Since maize requires a lot of nutrients and moisture, these two types of gardens were suited to maize cultivation. In the *mateshi* gardens, maize benefited from nutrients resulting from burning, as well as from heavier soils which retain moisture more efficiently. However, during prolonged droughts, the *mateshi* maize perished with the finger millet.

The subsequent expansion of maize cultivation in Kaputa district is linked to broader national trends. The post-colonial government policy was geared towards the wholesale promotion of maize production. The state controlled agricultural support institutions, such as marketing co-operatives, input supply, credit schemes, agricultural extension and research, etc, which were reoriented towards the promotion of maize as a cash crop. Although the new hybrid seeds promoted by the Department of Agriculture do not flourish in Kaputa owing to the harsh climatic conditions, there is evidence that the more hardy, open-pollinated flint varieties are being grown on a much larger scale than ever before.

The increased significance of maize in the local economy was accompanied by strategies to adapt the crop to local environmental and socio-economic conditions. The decline of *mateshi* cultivation has meant that all crops grown with or after finger millet in *mateshi* plots had to be grown elsewhere (or abandoned). Farmers have evolved strategies to utilize more efficiently the *chipya* for crops other than cassava, such as maize. This represented a technical challenge for the local people because *chipya* is less suitable for maize under conditions of erratic rains, which led to the development of management strategies to conserve moisture and enhance soil fertility.

In other parts of the province, maize is grown on the flat in straight lines or on thin and low ridges. In *chipya* soils, farmers tend to grow maize on mounds which are unusually large, with several plants randomly scattered on the mound. These mounds are often very thinly planted, presumably to reduce competition for limited moisture (see photo 18). By contrast, maize grown on *dambo* margins could be planted on the flat, or on thin and low ridges, as elsewhere in the province.

Traditionally, the primary aim of mounding is to improve soil fertility for root crops such as cassava and sweet potatoes through the decomposition of incorporated organic plant material. Under such conditions, the aspect of moisture conservation could simply be considered an unexpected benefit. The huge mounds created for growing maize in Kaputa serve the dual purposes of enhancing the fertility of depleted *chipya* as well as counteracting the effects of drought. Local farmers also maintain that it is less labour intensive to make and weed large mounds, than it is to produce maize on the flat.

Middle-aged and old households are more likely to be involved in farming because they are less involved in fishing. This gives them greater opportunity to diversify the range of crops grown. In addition, such households tend to include a larger proportion of retirees from urban employment who have investible income to engage in maize cash cropping. Both men and women are involved in seed-bed preparation, but subsequent operations, such as weeding, harvesting and processing, are done by women.

The level of market orientation seems to influence gender divisions of labour in the maize enterprise. For more market-oriented households, maize growing is treated as an economic enterprise under the control of the male household head. In such circumstances, men tend to participate to a much greater degree in most of the operations, from land preparation to harvesting, threshing and bagging. They may also assist female members of their households by hiring paid female labour to perform tasks that are regarded as feminine, such as weeding and threshing.

Dambo utilization

Dambo utilization is an additional strategy to use limited water resources more efficiently in adverse climatic conditions. Unlike other indigenous SWC techniques which attempt to 'harvest' or conserve water, this strategy involves locating crops where there is moisture. This distinction is important because access to this technology is not only influenced by the efforts and disposition of individual households, but also by the availability of suitable cultivation sites. Crops in *dambos* are constantly changed to suit fluctuating flooding regimes, depending on seasonal expansion or shrinkage of the lake.

Before expansion of the lake, *dambos* were used to cultivate crops such as rice and maize. By the mid-1980s, some 20 per cent of the households in the area were growing rice (ARPT, Northern Province, 1986). However, by 1994 no one was growing rice any longer in the study area owing to exceptionally poor rainfall during the previous two seasons. Instead of rice, there could be found maize, 'modern' vegetables such as cabbage, rape and onion, and sugarcane, bananas and even groundnuts. These upland crops are grown in the *dambos* to take advantage of soil moisture and to avoid the risk of drought in the upland areas. Thus, seasonal fluctuation, rather than long-term trends, characterizes cropping strategies on *dambos* in the study area.

The degree of the participation of different social categories in *dambo* cultivation is dependent on the range of crops being grown in a given season. During wet years, when the *dambo* is used for rice cultivation, women tend to participate more because rice is treated exclusively as a woman's crop. This is partly because at the time of land preparation for rice, men were involved in fishing and in preparing fields for cassava (ARPT, Northern Province, 1987). During drier years, when crops such as maize, groundnuts and sugarcane are grown, both men and women participate.

CONCLUSION

Ecological and demographic pressures have led to new modes of adaptation in the lakes depression zone. Local people are continually adjusting their practices to deal with the challenges of the time. External intervention must build upon what the local people are already doing, instead of seeking imported solutions. Interventions should be aimed at developing a flexible basket of options to suit the ever-changing production opportunities and constraints for different categories of people in the community.

As shown, the shift to *chipya* soils for cassava cultivation has brought new challenges for farmers, for which new solutions should be sought. Since the *chipya* soils are less fertile, existing cassava varieties take longer to mature than was the case under the *mateshi* system. The development challenge is thus to identify ways to enhance the fertility of *chipya* soils without eroding the low-labour input advantage of cassava which has made it a feasible option for households which are engaged in off-farm activities such as fishing. Similarly, interventions should not exacerbate the labour burden on women who already undertake most of the operations involved with cassava production.

The increased significance of maize as a complementary starch staple in place of finger millet has also been noted. The fact that people take the trouble to make large mounds for maize cultivation indicates the importance which is attached to maize production as a complementary livelihood option. There is scope, perhaps, to make this strategy more remunerative by optimizing yields per unit area. Another area of development is the possibility of intensifying and expanding the use of *dambos*. As already noted, the strategies used by the local people to utilize *dambos* are often very complex and require more understanding by outsiders.

SWC is only one of a number of adaptation strategies to enhance agricultural productivity under difficult environmental conditions for the ultimate purpose of securing a livelihood. In many cases, agricultural production may not be the most remunerative adaptation strategy. If this is so, the success of indigenous SWC will depend on how well it measures up to other options in the basket of adaptation strategies which are available to the community.

Fishing as an adaptation strategy has a direct bearing on the indigenous SWC strategies which are used in farming because agriculture and fishing both complement and compete with each other. Any SWC intervention which disrupts the complementarity between fishing and cassava cultivation may not be acceptable to the local community. Fishing is the most important source of income in the lakes depression zone. However, cassava cultivation occupies a central role in providing the caloric requirements for the community. According to locals, most people who farm, fish as well. This gives them an income and allows them to remain self-sufficient in starch staples at the same time. Less labour-intensive crops such as cassava, as well as new less labour-intensive methods of fishing brought about by its commercialization, allow a suitable balance to be struck between farming and fishing.

As this paper has illustrated, the expansion and contraction of the major vegetation and land systems in Kaputa district is a major determinant of the modes of land use. Most of the SWC strategies identified in this study are local adaptations to these changes. Against this backdrop, SWC should be seen as a dynamic and continually unfolding process arising from concrete historical experiences, rather than as a set of 'techniques' which can be scientifically deployed or disseminated wholesale.

14

'GRANDFATHER'S WAY OF DOING'

Gender relations and the *yaba-itgo* system in Upper East Region, Ghana

David Millar with Roy Ayariga and Ben Anamoh

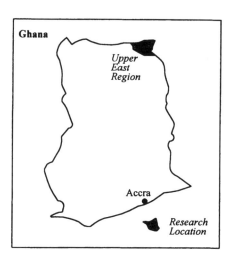

Many studies of indigenous soil and water conservation (SWC) downplay the role of women in the generation, use and perpetuation of locally appropriate technologies. This study of the *yaba-itgo* system of cultivation practised by the Frafra in the Upper East Region of Ghana demonstrates that simplistic assumptions about male farmers being the sole or key players in these conservation systems do not adequately explain reality.

INTRODUCTION TO THE AREA

The Upper East Region, located in the savannah grassland belt, is bordered to the north by Burkina Faso and to the east by Togo. Population density in the Frafra area is high at 204 persons per km^2 (IFAD, 1990, p 22). Vegetative growth is mainly of grasses and shrubs, with some scattered trees. The region experiences distinct rainy and dry seasons, with the rains beginning in May and ending in September. Average rainfall is 800–900mm, but is unevenly distributed and extremely erratic in onset, duration and intensity. A lull in rains during the mid-season is not uncommon. The area is affected by dry and dust-laden Harmattan winds during the dry season.

In the study area, Zanlergu, many soils are of low intrinsic fertility, with characteristically low organic matter content and nutrient retentiveness. In combination with sandy textured surface horizons, low organic matter content results in a low moisture holding capacity of soil in the root zone. This can be a serious constraint to cultivation because dry periods tend to occur within the growing season, usually in June or July.

Agricultural and livestock production

Agriculture in Zanlergu is mainly rainfed; during the dry season, cropping ceases except in small garden areas located along river-beds or dams. Most farmers practise mixed cropping on holdings of an average size of 3ha. Sorghum, maize, yam, cassava and beans are grown alongside the main staple millet. Cultivation is mainly subsistence-oriented, although rice, cotton and groundnuts may be grown on a small scale as cash crops. Continuous cropping is the norm, although a few farmers use a two-year bush fallow system. Most farm households in the area keep small-stock such as sheep, poultry and goats, and, in some cases, cattle. Livestock function as an important insurance substitute during lean times.

Most farmers fail to generate sufficient income for reinvestment in their enterprise, so farming systems remain highly dependent on simple hand equipment such as the traditional hoe, cutlass and sickle, and minimally dependent on external farm inputs. Only 5 per cent of the farmers use bullock traction; the high cost of equipment and animals, combined with the steep and stony topography in the area act as constraints to further uptake. On average, yields and net returns to resources employed in farming are relatively low.

Labour allocation and gender divisions of labour

Extended farm families are common in Zanlergu; up to 20 households may reside in a single compound. Goods and services are both produced and consumed within households, although systems of exchange exist at the inter-household level and may be drawn upon when labour or other resources are short.

Although farms are run as household enterprises, very clear divisions by gender and age with regard to tasks and responsibilities characterize farm operations (see Oppong, 1973, for an early study). However, where household composition has changed, whether through migration, separation or widow-hood, these differences tend to be less distinct. In general, men prepare the land, while responsibility for sowing and first weeding falls to women. Men under-take the second weeding, and women the harvesting and processing. Deviations from these gender-specific roles are not uncommon, in which case men and women may work together on specific farm activities. Divisions of labour sometimes depend on crop type, with different sets of responsibilities associated with cereals, tubers or vegetables. Produce from family farm plots is kept as food for the household, whereas produce from individual plots is kept by the cultivator, whether male or female, to cover hungry months, or sold for cash. Men mostly grow millet and women produce groundnuts, rice and vegetables on their personal plots. Some women have a few chickens and, occasionally, a goat or sheep. Overall, women assume control over the production, processing and marketing decisions for crops grown on plots allocated to them by men.

Responsibility for labour contributions falls heaviest on the young men and women; older men and women with sons and daughters to work for them often assume an advisory role over others' work, and tend to contribute labour only where highly skilled tasks require their attention.

Household farm operations are planned during discussions or 'joint' deci-sion-making between a male head of household and his wife. Women are often charged with responsibility for keeping a mental schedule of the agreed tasks (Millar, 1992). However, as male household heads normally have the over-riding decision-making power over farming operations, or, for that matter, over other issues relating to the family, such as marriage, other cultural events and practices, or the use of family assets, women's control over decision-making clearly does not equal that of men.

Land tenure and access

Land holdings are small, often not exceeding 2ha per household. Bush and compound lands may be controlled individually or collectively while individual and family land is commonly privately owned. Although land held as family property is rarely sold, it is frequently leased. Tenure insecurity in Zanlergu is high, as can be seen by the high incidence of land disputes. Where land is rented, the ultimate right of repossession and, more immediately, control over trees and other perennial features on the land, remain under the control of the original owner. This acts as a severe disincentive to investment in land. Furthermore, some landowners prefer to rotate the use of the land they lease out each year, which again works against appropriate land management or investments which would improve productivity in the long-term by those leasing land.

Gender-specific insecurities cross-cut those faced by tenants. Under indi-genous tenure arrangements, control over land is vested in a male household head. Within this system, women's rights to land are secondary and contingent on their relationship to male kin. Although women may have rights of access to land, their overall control over land-use decisions and output is often constrained by the landowner. This may prevent women investing in soil and water conservation measures on land which they have been allocated, or deter them from managing appropriately land for which they are expected to contribute labour. For instance, women interviewed during the research argued that groundnut cover protects topsoil from rainwater impact and improves soil fertility. Following the harvest, the vines are left to rot *in situ*, leading to soil enrichment. Yet women seldom reap the benefits of these soil improvements in the long-term because men tend to appropriate this enriched land to grow their millet. This system explains why outputs of 2000kg/ha of millet are common from the seemingly bare and poor soils in Zanlergu. Men justify their actions by arguing that this is a rotational system which maximizes the benefits to the soil. Yet it is arguably an exploitative one.

External interventions

Infrastructure is poor in the study area. Many farmers are cut off during the rainy season because of impassable roads and are therefore unable to sell or buy at prevailing market rates. Middlemen take advantage of this to buy produce at relatively low prices and to sell on at significant profits. Low prices for farm produce, despite the high cost of production, create a big disincentive to cultivation (Runge-Metzger, 1988). The poor condition of roads also means that farm inputs fail to reach farmers at the right time and in good condition.

There is some exposure to outside agencies in Zanlergu: agroforestry, alley cropping and live fencing technologies are currently being introduced. However, because of the transfer of the technology model used to disseminate information on these technologies, farmers tend to abandon them once the external force or motivating factor is withdrawn.

In summary, Zanlergu is an area of poor natural resource endowment for agriculture, where returns to labour and inputs are generally low, and external technical intervention to date has largely failed to make a significant impact on farming systems. Tenure insecurity is a fact of life for those who rent land from better-endowed farmers, and is rife among women who hold only secondary and contingent rights to land. Men's control over women's labour for the cultivation of family fields means that their work burdens are substantial, although later on in the life-cycle, these burdens may be eased when sons or daughters are old enough to be delegated work.

YABA-ITGO: 'GRANDFATHER'S WAY OF DOING'

Yaba-itgo, meaning 'grandfather's way of doing', is used to describe a wide range of conservation techniques. In an area where average precipitation is 800–

900mm and where slopes are steep, the main aim of this system is to conserve moisture. However, in a few cases some techniques are used to encourage run-off as a way of avoiding flooding in low-lying areas. The techniques are never used in isolation, but in combination as integrated parts of the farming system. Slope percentage determines both the choice of which measures to use in combination and the intensity of use.

Gender roles and labour input into erosion control

The majority of soil erosion techniques are highly dependent on household labour availability, and less so on communal labour or labour hired from within the community. A breakdown by gender of labour inputs into conservation work (see the table below) shows that women's input is substantial and exceeds

Breakdown by gender of labour inputs into conservation work

Practice	Age	Indigenous SWC techniques	Major crop	Gender roles and labour inputs
Stone bunding (widely practised)	Very old	Contour stone lines set out in grid. Big stones placed at base, small ones used as fillers. Yearly maintained. Combined with vegetable production (women grow okra and kenaf along stone walls) and terraces. The root system of these plants holds together soil particles that settle around the stones, forming a mulch. Used on very steep, hilly land with loose sandy loam soils, which are prone to run-off.	Millet and groundnut	Men do C* (see key below). Initial construction (mainly stone collection) requires 20 working days per ha. Women do M* and R*. They are also responsible for arrangement of stones at construction stage. Maintenance and repairs required every year because of livestock damage. Requires 3.5 working days spread over a 3 month period. Children also contribute significant labour.
Contour tillage	Very old (new with bullocks)	Contour cultivation with hoe or plough. Slope direction is assessed by thickness of grass, water direction, collection of pebbles. Ridges of 0.25m width, 10m long and 0.5m apart. Useful where slopes are gentle.	Millet and sorghum	Men do C. Where men have access to bullocks, work takes two days per hectare. Using a simple hoe requires eight working days per ha. Women do M and R. Maintenance during the farming season takes six days per ha during the farming season. Women generally do not have access to bullocks. *Continued on next page*

Breakdown by gender of labour inputs into conservation work (*continued*)

Practice	Age	Indigenous SWC techniques	Major crop	Gender roles and labour inputs
Border grasses	Old	Grown to check water and wind erosion. Strips of various species 0.5m wide, densely grown across slopes. Most suitable on flat pieces of land. Often planted on plot boundaries. Grasses harvested and used as roofing, brooms, for weaving baskets, mats, etc. Also used as organic matter when dead.	Millet	Men do C. Women do M, C and R. Planting and care of grasses takes about 12 working days over the farming season.
Strip cropping	Old and new	Crops planted in alternate contour strips to control water and wind erosion. Legume-cereal or cereal-cereal combinations. Used when gradient not very steep and run-off is slow. Recent: alley cropping and agroforestry.	Millet and groundnut	Men do M, C and R. Women do M, C and R. Average input of 14 days per season.
Terracing (widely used)	Very old	Used on very steep slopes. Limited to cultivated areas. Variable dimensions. Steep back slopes of narrow terraces are most common. Terraces set out in grid. In combination with stone bunding and vegetable growing as border strips, run-off may be reduced significantly. This allows weathered material to remain in situ. Very rich surface soils around the stones result and vegetable cultivation can be highly productive.	Millet and groundnut	Men do C, M and R. Preliminary construction requires 15 days. Women do C, M and R. Maintenance requires 10 days per season. Children also contribute significant labour.

* C: construction; * M: maintenance; * R: repair.

that of men if hours contributed are totalled. Far from contributing supplementary labour as men's 'helpers', women hold primary responsibility for much of the maintenance and repair work which is required to preserve erosion control structures, and in many cases also they contribute to the initial construction work, although tasks may vary from those of men. For example, in stone bunding, men gather and women arrange the stones used in construction.

Crucially, much of the work undertaken by women is on-going, or repeated at specific intervals throughout the year, perhaps at the beginning or at the end of the cropping season, while much of men's work (ie, in bund construction, contour tillage and cultivation of border grasses) comprises one-off, time-bound tasks. A further difference lies in the fact that men are more likely than women to have access to bullock traction to assist them with their work, while women's tasks predominantly require the use of simple, labour-intensive technology. Children's labour input into *yaba-itgo*, managed by women, is also significant, and adds a further dimension to assumptions of men being the key players in soil and water conservation activities.

Gender roles and labour input into drainage control

Flooding is uncommon in Zanlergu, except in a few valley bottoms and flat lands. However, because rice and vegetable growing – cash-generating crops – are the major activities in these low-lying areas, local people consider investment in drainage-related conservation activities worth the effort. Communal labour is drawn upon to a far greater degree than is the case with erosion-control technologies, and this is justified on the basis that drainage improvements on one plot may have a negative impact on neighbouring plots. The table below shows the breakdown by gender of labour inputs into drainage control.

Breakdown by gender of labour inputs into drainage control

Practice	Age	Indigenous SWC techniques	Major crop	Gender roles and labour inputs
Land smoothing/ levelling	Very old	Elimination of low points by movement of earth to improve drainage. Direction of flow is by visualization. Used on semi-flat lands. The overall aim is to combine even distribution of moisture on farm land with drainage improvements.	Groundnut, rice and vegetables	Men do C*, M* and R*. Women do C, M and R. Requires 12 days to level 1ha using simple hoe. Regular maintenance needed to repair washed-off areas.
Graded furrows	Old	Excess water disposal with minimum erosion. Furrows 0.5m wide. Used in valleys. Furrows partly filled with stone to slow down flow. Stone dams may also be constructed at downstream end of furrow where water flow is expected to be high.	Rice land and vegetable gardens	Men do M, C and R. Initial construction of furrow requires four days per ha. Women do M and R. They also complete initial design by arranging stones. Repair and maintenance requires input of 10 days per ha over season. Simple hoe technology used by both men and women.

* C: construction; * M: maintenance; * R: repair.

Rationale and perceptions about techniques

In *yaba-itgo*, key objectives of water harvesting and conservation, soil conservation, and drainage and fertility improvements overlap as operational activities and also in terms of intended benefits. Women perceive the conservation technologies within *yaba-itgo* to be an important way of satisfying their ancestors, although this customary belief is gradually weakening. In terms of the present, production benefits provide the strongest driving force. Nevertheless, the Frafra also associate *yaba-itgo* with investment in the long-term sustainability of the land – in other words, as a means of preserving resources for their descendants. Today, high population densities and poverty intersect to give local people an impetus for conserving the resource base. This has found expression as renewed interest in preserving sacred lands, shrines and groves, and in restricting bush burning and bush-fires in the locality, as well as affecting the nature and extent of the use of *yaba-itgo*.

People in Zanlergu recognize that land use and land management techniques ensuring the sustained and continuous use of limited cultivatable land, and providing increased yields to feed a growing population, have to be evolved and maintained at minimal operation costs. To this end, three major changes have affected *yaba-itgo* in recent years. First, soil and water conservation technologies have been blended with agronomic practices, such as mixed cropping, cultivation of groundnuts and mulching with crop residue and grasses. Secondly, farmers have made a conscious decision to cease bush burning with the aim of regenerating organic material which, in turn, will improve soil fertility. Thirdly, new technologies such as live fencing and the use of bullocks to plough across slopes have been incorporated to some extent into the system. However, the prohibitive cost of bullock traction and the fact that it is only adopted for activities such as ploughing which are specific to men within the gender division of labour, means that women tend to resist it.

CONCLUSION

Investment in soil and water conservation depends to some extent on how much labour people have at their disposal. The labour-rich are capable of meeting the intensive labour demands of these practices far better than those facing labour scarcities. It also depends on people's incentives for sustaining the land, which, in turn, are closely associated with security of tenure. Women in general and tenants in particular both cultivate land under temporary arrangements and expect land to be taken away from them at short notice. Under these conditions, people may perceive investment in long-term land improvements as inappropriate because they are unlikely to reap the benefits of their work.

In Zanlergu, women are widely considered to be more meticulous in ongoing, repetitive maintenance and repair work than men. Local people assume that the more female labour is available to each household, the more efficient will be the *yaba-itgo*. However, the large gap between the labour commitments expected of women under this system and the unlikelihood that they will benefit from their work means that the long-term sustainability of *yaba-itgo* itself is in question. Building up women's incentives for conservation work, not least by improving their rights of control over land, needs to be central to all policy intervention in this area.

There are more general issues at stake. If technologies are introduced to the area, serious consideration needs to be given to their cost. A credit system which is built in to support their adoption may be needed, and access by women on equal terms with men needs to be promoted. This is especially important because women often have access to far smaller cash resources than men. The mode of introducing technologies is also at issue. Top-down, transfer of technology models need to give way to a participatory approach because different interests, whether by gender or by other social attributes, are more likely to receive attention when local people participate in technology selection and introduction. To date, innovations introduced by Zanlergu from outside have more often than not had a limited impact because they have been aimed at the wrong people. Recognizing the part played by women in cultivation and conservation work, and gender differences in interests in and incentives for this work, is a first step towards designing interventions which will build on and sustain *yaba-itgo* into the future.

15

HOW RICE CULTIVATION BECAME AN 'INDIGENOUS' FARMING PRACTICE IN MASWA DISTRICT, TANZANIA

J M Shaka, J A Ngailo and J M Wickama

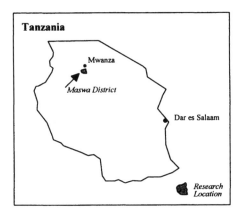

Water conservation techniques have played a critical role in the development of smallholder rice cultivation in the Maswa district of Tanzania. Rice cultivation cannot be said to be an 'indigenous' practice, yet it now has the status of a 'traditional' crop despite the fact that it is only about 40 years old. How and why was rice introduced to this region?

This paper traces the shift from cotton to rice cultivation in Maswa district which brought about corresponding significant changes in land use. Farmers began to experiment with water harvesting techniques and to adapt them to local conditions, largely replacing cotton as the main cash crop. The remaining high-market value of rice induces farmers to continue to invest in extensive soil- and water-conservation (SWC) measures within whole catchment areas to make sure that all water is utilized efficiently by all farmers living in the area.

INTRODUCTION TO THE PEOPLE AND THE REGION

The Maswa district of the Shinyanga region of Tanzania has few major roads and accessibility within the district is difficult during the rainy season. The railway traversing the district in the west is vital for transporting goods, including cotton and rice, to Dar-es-Salaam or Mwanza.

Maswa district is part of an upland plain which consists of gentle slopes. The central part forms a more or less continuous ridge, interrupted by wide valley bottoms and plains. The ridge runs in a north-west direction; in the south-eastern part it ends in a broad, low-lying plain with numerous depressions, once a lake-bed. The depressions provide important sources of water during the dry season. Granitic hills are a common feature. The scarcity of year-round flowing rivers means that sources of permanent water are largely absent.

Ethnically, the rural population in this region are almost all Wasukuma. Tsetse fly-control measures which began in the 1920s, together with the eradication of rinderpest through large-scale immunization programmes in the 1960s, opened up the region for agriculture and pastoralism. The population of Maswa district has continued to expand substantially over the last decade. Families have migrated from the less fertile region to the north of the district, swelling the population density from around 40 people per km^2 in the 1960s to the current density of between 50 and 70 people per km^2.

The importance of cattle

The Wasukuma keep large herds of cattle. Grazing land is more plentiful in the north of the district, which receives on average 900mm of rain a year, compared to only 600mm in the drier southern region. Farmers are forced to migrate to the northern region in search of grazing land, particularly between May and October when there is very little rain. Most of the rain falls between November and April, creating a short growing season of around 119 days. Depending on the amount of rainfall received, many parts of the district develop springs and mini-lakes during the rainy season. June to mid-September, on the other hand, is characterized by intense radiation and high evaporation, with clear, cloudless skies and limited plant growth owing to moisture stress.

Social differentiation is expressed in terms of cattle ownership and access to oxen and a plough. A farmer who has many cattle and owns an ox-plough, for example, is considered to be rich, because he or she is able to cultivate relatively big rice-fields and other crops and practise water harvesting techniques in his or her rice-fields. Farmers with fewer or no cattle depend on borrowing a plough or using hand-hoes and usually cultivate less land.

Cattle are also the most important means of investing money, as agricultural surplus is normally converted into cattle. They are considered to be a store of wealth that entails security, social status and power. The exchange of cattle as bride-price is an important aspect of marriage. At the same time, oxen are the backbone of the farming system. More than half of the farmers use ox-ploughs for land preparation. Ox-carts are also widely used to transport agricultural products, firewood and water. Animal dung is also used as a source of fuel during shortages of firewood, although the use of manure as fertilizer is limited owing to the lack of transport (Tanzania/Netherlands Farming Systems Research Project, 1989).

Using and owning land

The first areas to be farmed in the Maswa district were the hilltops and ridges as their lighter soils were easier to work. As these were exhausted, farmers were forced to use the valley slopes and finally the valley bottoms, areas with heavier soils which in some cases were susceptible to waterlogging.

Soil types vary throughout the region according to the topography and underlying geology. The Wasukuma have names for a number of distinctive soils that occur in the area (Milne, 1947). This indigenous terminology classifies soils according to their suitability for cultivation of various crops which often depends on their position on the landscape and are based mainly on soil depth, workability and susceptibility to waterlogging (Milne, 1947). Their language contains a richer nomenclature for soils than that of any other ethnic group in East Africa, and their soil classification system corresponds closely with the Western concept of soil sequences (Williams and Eades, 1939). The dominant soils with their broad characteristics and parent material are summarized in the following table.

Land tenure in Maswa district is based on rights of occupation and customary land ownership. Land plots which are owned by the village are allocated to households according to land availability. Once allocated, these plots remain in the hands of the user, who maintains the right to cultivate until he or she decides to migrate to another village. Their right to use the land is passed on to their children through the husband's lineage.

In cases where an individual does not own land, they are able to rent or borrow land from villagers who have enough, either short term (one season) or long term (several seasons). The use rights of land renters or borrowers are more insecure than those of people who own the land. They are also less likely to invest in long-term development, such as water harvesting techniques. By contrast, those with secure land rights are more likely to invest in SWC.

SOIL AND WATER CONSERVATION ACROSS THE LANDSCAPE

Historical shifts in land use

As in many parts of Africa, land use in Maswa district was based on shifting cultivation. Initially, the farming system was based entirely on the cultivation of food crops. Until the 1950s, the staple food was bulrush millet (mainly in the north) and sorghum (mainly in the south). Since then, these crops have been

Local soil classification

Local name	Luseni	Itogoro	Mbuga
Description	Well-drained sandy soils, yellowish-brown, loose structure	Imperfectly drained, loamy to clayey soils, dark grey	Poorly drained, cracking clays and heavy clays, black
Location	Ridge crests, upper slopes	Middle and lower slopes	Valley bottoms
Characteristics	Low capacity to store water, easy to work, good infiltration, topsoil potentially rich in organic matter but prone to leaching	Calcareous, difficult water infiltration owing to the more clayey subsoil, moderately fertile, may be subject to heavy run-off	Calcareous, very narrow moisture range, highly fertile, difficult to work, prone to waterlogging
Crops	Groundnuts, sweet potatoes, cassava, sometimes maize and cotton	Intensive rice cultivation (where there is waterlogging), cotton, maize and cow peas elsewhere	Communal grazing, some cotton and sorghum in areas with improved drainage conditions
FAO soil classification	Eutric Regosols, petroferric phase	Calcaric Phaeozems, sodic phase	Calcic Vertisols, sodic phase

replaced largely by maize, although sorghum is still grown in some places as a drought-resistant crop (Collinson, 1972).

The end of World War II brought significant changes in land use. Shifting cultivation was replaced by fallow and permanent systems owing to increases in population density and the introduction of ploughs (Rotenhan, 1966). Cotton, which had been introduced by the German administration at the beginning of the century, increased its acreage during the 1950s, and Maswa became a major cotton-growing area which supplied the export market until the 1970s when cotton production steadily declined after producer prices dropped and market conditions deteriorated. Cotton has been replaced since by rice as the main cash crop. It has no market restrictions and is grown using well-established water harvesting techniques. On average, around one-third of the total area is now reserved for rice cultivation, and farmers spend up to half of their time on this crop.

Harvesting water for lowland rice

Why have water harvesting techniques become so important in Maswa? Rice cultivation was not entirely new to the region: as early as the 1930s, a small number of Asians working in the local cotton ginnery grew enough rice for their own purposes. A food shortage during 1949 encouraged some local people to experiment with rice cultivation in valley bottoms and in depressions which held stagnating water. No bunds were used in these early rice-fields: a farmer simply had to locate an area with suitable soil with water-retaining abilities. The depressions on the valley floor were the best areas for this kind of rice cultivation, given the waterlogged conditions of the soil.

As more people realized the importance of the crop, suitable depressions and valleys became more scarce. Rice cultivation started to shift to the valley sides where techniques for collecting water as run-off were developed. In order to collect run-off water from the higher slopes, small bunds were constructed around the fields. Initially, round ridges called *migoka* were developed, which later became rectangular or square, depending on the size of the farm and gradient of the slope. As the population pressure increased, many more people began to experiment with rice cultivation and water harvesting techniques.

The main function of the bunds is to keep water in the rice-fields. Other important functions of the bunds are to demarcate rice-fields and to act as paths between the fields. They are made of soil and strengthened by being planted with grasses such as Guatemala grass (*Tripsacum laxum*) and *Hyparrhenia* spp.

Methods of construction and maintenance

The construction of rice bunds is both time consuming and labour intensive. This activity normally takes place towards the beginning or end of the wet season when the soils are still wet or moist and can be tilled easily. The time used to make bunds varies according to the size of the field and the labour-force available. The main sources of labour on which farmers draw are:

- family labour;
- hired labour;
- traditional groups;
- neighbours and friends;
- draught animals.

Family members – ie father and sons – are the most dependable labour-force. A big family is able to cultivate a larger area. Very rarely and only in critical times are the female members of the family involved in bund construction.

If family labour is not enough or there is a need for the family to increase acreage, other labour alternatives are considered. If a farmer has a good relationship with his neighbours, he can ask them to assist him. Friends and neighbours are paid in kind and not in cash. Hired labour is usually expensive and therefore is not used often, although it becomes cheaper during drought years. Labourers may come from within the village or from distant villages. Traditional work parties are groups of more than 10, or even 40 people who have decided to work together and are relatively cheap to employ. Tractors are used by very few farmers in this area as they are too expensive. Also, even when the bunds are made by tractors, the finishing (including levelling) still has to be done by farmers themselves using a hand-hoe.

Draught animals are perhaps the most important labour-force. They are used most intensively in preparing the middle section of the fields. A farmer who does not own cattle can use neighbours' or friends' oxen and pays by providing the ox-owner with a place to graze. In some cases, ox-plough owners team up to prepare the land before the fields are flooded. Bunds made with the help of oxen are small and may also need to be reinforced using human labour.

Another aspect of water harvesting which needs labour is the levelling of fields in order to encourage the even distribution of run-off water within the field. This is a labour-intensive task using only the hand-hoe and spade. Farmers prepare their fields before the rains and level it during the wet season so that they can see how water is distributed in the field.

Gender aspects of rice cultivation

Gender divisions of labour are based on the traditional norms of the Wasu-kuma. The father is normally solely responsible for rice cultivation, but will mobilize the male family members or other work-group parties to prepare the rice bunds. The mother ensures that the family has other sources of food crops, cultivating sweet potatoes in fields which are considered her responsibility. To some extent, this increases the labour burden on women, particularly when the family has no male members to cultivate the rice-fields. Women will be expected to participate in bund preparation when the family has no male members to cultivate the rice-fields, in addition to maintaining their own fields. Sowing, weeding and the harvesting of rice, on the other hand, is done by both male and female members of the family.

The maintenance of rice-fields, particularly the bunds, is done throughout the growing season to ensure that water stays in the fields. It is the father's responsibility to sell the produce in the market, although in families with no male head of the household, the mother will do it herself.

Using the catchment area to encourage run-off

In order to achieve maximum water harvesting, the mid-slope soils (*itogoro*) are kept bare by over-grazing. These over-grazed lands are purposely left by farmers for their livestock to graze during the dry season. With the first rains, these soils develop a surface crust which reduces infiltration so that maximum water run-off reaches the rice-fields below. These bare-slope segments (or catchment intervals) are made large enough to ensure that a large amount of rainfall runs off and can be collected in the rice-fields downslope through openings in the bunds. In places, shallow channels are made to lead the water to these openings. The catchment intervals are never too long, so that rill and gully erosion is avoided and no damage is done to the rice-fields. On very long slopes, farmers construct two or more segments of bunded rice-fields separated by water-catchment intervals, thus controlling erosion. Farmers have learnt through trial and error to match the size of their fields with the size of the catchments above them, always trying to ensure that there is adequate run-off water which can be harvested in the rice-fields. The regeneration of the vegetation on the grazed catchments is slow, but takes place each season and therefore does not worry the farmers in the short term. In some cases, the catchment ridge is also cultivated when land for other crops is scarce. These ridges are placed longitudinally down the slope to encourage run-off to flow into the rice-fields.

Gullied fields may be rehabilitated because the bunds are always put across the gully or waterways. To reduce gully formations further, farmers control rills and small gullies in their fields with pegs across the flow.

These SWC methods, which have been developed over the last 40 years, are not without some disadvantage. As described above, the demand for labour is very high, particularly for the construction and maintenance of the bunds and for levelling. The extra water means that more time is needed to be spent on weeding. Lastly, available grazing land is reduced, and in some cases roads are destroyed in order to make sure that run-off from roads leads directly into a rice-field. Farmers, however, will not necessarily see these as problems or disadvantages because they are done to increase the yield of the rice-field.

CONCLUSION

The introduction of wetland rice into the area brought about significant changes in land use, together with improvements in food security. Cotton was grown on the sandy soils of the upper valley slopes, together with maize, the main food crop. Maize yields had never been high because of the low, erratic rainfall and the adverse soil conditions. Rice thus provided a much needed diversification of crops which, unlike cotton, could be used both as food and as a cash crop. It also offered the possibility of exchanging food directly (rice for maize). Another advantage of rice cultivation is that recurring investments to grow the crop are much lower than those for cotton. The latter needs pesticides and hence renders farmers more dependent on the external material inputs. Bund preparation, on the other hand, although labour intensive, can be undertaken by any farmer, although in practice those with relatively large families who own oxen are more likely to invest in water harvesting techniques than poorer families with no livestock.

Unlike cotton and maize, wetland rice cultivation offers the possibility of retaining surplus water at low cost in the bunded fields. Before the introduction of rice, cotton was planted mostly on the slope, so that excess water could escape during heavy downpours, thus avoiding waterlogging. Today, in parts where cotton is still grown, ridge cultivation often still holds the slope, with the aim of providing water to the rice-fields located further downslope. In this way, surface water and eroded soil (containing plant nutrients) are retained in the system. No water and soil are lost to the *mbuga* soils in the valley bottoms. The *mbuga* are now rarely waterlogged as rice-fields on the bordering lower slopes catch any excess water. As a result, more *mbuga* soils are now being cultivated in places with cotton and sorghum in broad beds because the risk of flooding has been substantially reduced. Water reservoirs situated in the *mbuga* no longer silt up as quickly as in the past.

Not surprisingly, more and more farmers are investing their labour, time and money into water harvesting techniques for rice cultivation in Maswa district. The economic returns from rice cultivation are much higher than for other cash crops, and it enjoys a higher price in the official and free market. The expected yield of rice per acre of a treated field is 15–20 bags of husked rice.

To achieve a successful all-round cropping system, a farmer tries to have a continuous strip of land, from the ridge crest to valley bottom, so that he or she is able to grow a diversity of crops on the various soils. Maize, groundnuts, sweet potatoes, cassava and cotton are grown on the ridge crest and upper slopes, while rice is grown on the middle and lower slopes in *itogoro* soils. *Mbuga* soils on the valley bottom are maintained mostly as communal lands for grazing cattle. The indigenously developed land use system in Maswa district thus currently fulfils two important functions for the Wasukuma: the need for food security and the opportunity to benefit economically from rice cultivation.

16

MAKING THE MOST OF COMPOST

A look at *wafipa* mounds in Tanzania

A C Mbegu

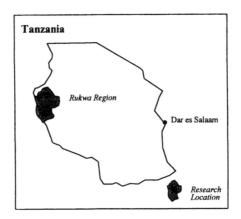

Wafipa mound cultivation has existed as a traditional subsistence farming method in south-west Tanzania for longer than there are written records. Early travellers and missionaries passing through the Ufipa Plateau in the 1880s described the system as 'primitive, inefficient, and wasteful of forest resources'. As a result, successive government policies have neglected the practice almost completely. No attempt has been made to study this simple indigenous conservation technique, let alone to improve the practice. As a result, the farmers themselves have widely forgotten the methods of *wafipa* mound cultivation which their forefathers formerly depended on. How do the farmers themselves perceive this indigenous farming method? This chapter attempts to look at the system through their eyes and, by field visits to numerous villages on the Ufipa Plateau, to determine the current extent of *wafipa* mound cultivation.

INTRODUCTION TO THE AREA AND PEOPLE

The Ufipa Plateau is bordered by Zaire to the west and Zambia to the south and forms the extreme south-west part of Tanzania. Situated between lakes Rukwa and Tanganyika, the plateau forms a high (1600–2000m) undulating plain of hills running from north to south, extending for about 225km in length and for 50–80km from east to west. Most of the Ufipa Plateau is devoid of woody vegetation. The entire area receives an average of 900–1000mm of rainfall annually during the wet season from October/November to March/April.

There are now approximately 443,000 people living in 214 villages on the plateau, making it the most densely populated part of the Rukwa region, 30 people per km², as compared to the region's average of 13 people per km². The national villagization programme in the 1970s resulted in a large number of 'new' villages being set up. Many of those living in these new villages are from the Wafipa ethnic group who have continued to live on the Ufipa Plateau and to cultivate the land. In spite of these new villages, the mode of living together in widely spaced but concentrated villages is traditional, and outmigration to urban centres seemingly remains low; only 4 per cent of those involved in this study said that any of their family members were in search of employment elsewhere.

The soils of the Ufipa Plateau are of poor quality, having developed from granite and gneissic parent material with very low clay and organic matter. They are therefore also low in both exchange and water-holding capacity. Given that the Wafipa are known to have lived on and worked these poor soils for well over a century, what is it that has enabled them to do so?

THE ORIGIN AND EVOLUTION OF *WAFIPA* MOUND CULTIVATION

According to early European travellers, the Fipa were strongly agricultural, and did not take part in warfare or hunting. Discussions with farmers revealed that the Wafipa have known for a long time about the types of soils in their area, and have consequently evolved efficient methods of fertility management, as well as soil and water conservation. The Wafipa traditionally recognize the soils in the following table.

In order to make the most of the relatively infertile soils on the plateau, farmers have developed a local method of making compost mounds, known locally as *intuumba*.

Making *wafipa* mounds

The farmer selects a site for mound cultivation on the basis of vegetation covering the area. Preference is given to an area which is covered with the tall grass known locally as *kasanza*, mixed if possible with bushes or acacia trees. The denser the vegetation, the better.

Wafipa soil classification

Local name	Description	Use
Chisenga	Sandy	Not preferred for cultivation
Mbulo	Blackish clay soils found around swamps	Sometimes used for growing sugarcane and sweet potatoes, but most often used for pot- and brick-making, as well as to plaster houses as decoration
Chipau	Whitish soils (rich in calcium bicarbonate?) found on hillsides	Used for decorative plastering of houses
Lusumba	Soils rich in plant nutrients which are usually found on termite mounds	Preferred for crop growing and making bricks
Chisenga and mubulo	Loam (consisting of sandy and clayey soils)	Preferred for crop growing
Itandala	Hard-pan, very common on the plateau (indicated by stunted plant growth)	Not chosen for cultivation

In March to early April both new and semi-decomposed old grass, together with small bushes, are slashed and put into oval-shaped piles and set on fire. The spaces consequently formed and covered with ash are then planted with *cucurbits*. In the rest of the area, mounds of around 90cm high and wide are constructed in rows spaced about 30cm apart by piling turf into mounds with the attached grass facing inwards.

Once the mounds have been built, they are planted with a leguminous crop, such as beans, cow peas, etc, or are left for the grass inside the mounds to decompose. Occasionally, cassava cuttings are planted at the base of the mounds. While the quick-growing beans/peas are harvested in June/July, the rest of the weeds are removed in October, towards the end of the dry season.

Most of the preparation work of the compost mounds is carried out by men. When the rains have started in late November/December, the mounds are broken down by the men and the whole area with its nutrient-enriched soil is sown with either millet or maize by the women. Once this crop has been harvested in June/July, small mounds are made over the weeds and residues left by the crop. These are then broken down again by the men in January, when the whole area is again sown with millet and maize. In the following rainy season, the whole plot is turned to ridge cultivation and planted with maize, beans, peas and other crops (except millet). After this, the plot is left fallow for four to ten years, depending on the abundance of land in a given area.

The cultivation of *intuumba* is difficult work. The Wafipa still use a special long hoe which used to be manufactured by local blacksmiths but which is now made by factories. It is hard, physical work, which young men or boys learn from about the age of 15. An average farmer can construct only 150–300 mounds per day, with a household cultivating a maximum of 1.6ha in a season. If a bigger area is cultivated, the household may ask for the help of neighbours, inviting the workers to drink the local brew, *chimpumu*, after work. Widows and old people in the villages have their mounds constructed for them using the village source of communal labour.

Current farmer perceptions of *intuumba* cultivation

While this method of cultivation has survived for centuries, our field visits to the area revealed that mound cultivation is fast disappearing as an indigenous soil and water conservation method.

First, and most importantly, well over 80 per cent of farmers who were approached have started to use animal-drawn ploughs on the Ufipa Plateau. A government drive which started in the early 1960s to introduce animal power in cultivation is having a negative impact on mound cultivation in the area. Most farmers are making use of the plough to compensate for lower yields by cultivating a larger area. At the same time, chemical fertilizers have been introduced, although they are only used by a small percentage of farmers who can afford them.

Most of the farmers who have started to use the plough have therefore abandoned the practice of *intuumba* cultivation. When asked why, the majority gave the reason that *intuumba* cultivation is much more labour intensive than the plough and therefore time-consuming.

Not surprisingly, field observations and discussions with farmers revealed that the traditional *intuumba* cultivation system is in decline. Only a small minority of farmers who cannot afford cattle, a plough and chemical fertilizers still practise the technique and, for those who do, most perceived the method to be in decline and did not see it as a method of cultivation which they could imagine still being in use in the future.

Current uses of *wafipa* mound cultivation

Despite the general decline in mound cultivation, it is still the preferred cultivation method for certain crops such as cassava and yams. Out of 80 farmers approached who now use ox-ploughs, 15 said that they still practise *intuumba* cultivation to grow tuber crops.

Composting is also widely viewed as a potential alternative to chemical fertilizers, the continuous use of which has been observed by many farmers to render the soils progressively less productive. Many farmers expressed a desire for *intuumba* cultivation if it were mechanized or otherwise made less labour intensive so that they no longer had to purchase expensive chemical fertilizers.

CONCLUSION

The compost mound technique has evolved from a practical understanding of the poor fertility of soils on the Ufipa Plateau. Such compost mounds are an indigenous use of green manure, not only to improve the fertility of the soil through the compost, but also in terms of the symbiotic effect of the leguminous plants which fix nitrogen and hence make the soil more fertile. In addition, the arrangement of the mounds in rows on slopes which are otherwise prone to soil erosion provides an important barrier to water run-off, thus further conserving both soil and water.

In the past, the Wafipa ethnic group living on the plateau were able to produce enough food for their own consumption, as well as some surplus which they sold to the gold-mining areas. Farmers today still believe strongly that without some form of fertilizer, either chemical, manure or as compost from the *intuumba* mounds, cultivation and crop production would be very poor.

In spite of nearly 30 years of a government drive towards introducing the ox-plough, it is pertinent that there is still a substantial number of farmers, even among those who have converted to the plough, who continue to rely on *intuumba* cultivation. In particular, tuber crops such as cassava and yam are frequently still grown on *intuumba* mounds.

In summary, it would seem that there is the potential to make *intuumba* or mound cultivation more relevant to the needs of farmers living on the Ufipa Plateau today. *Intuumba* cultivation incorporates composting and green manuring, which is still considered appropriate by the average smallholder. In the eyes of most farmers, the main disadvantage of compost mound cultivation seems to be the amount of time and labour which is required to construct these mounds. In order to capitalize on this good-will felt by farmers towards *intuumba* cultivation, it is important to bring it in line with current farming methods, in particular the plough, combined with an information programme on the additional benefits of *intuumba* mound cultivation.

17

CULTIVATING THE VALLEYS

Vinyungu farming in Tanzania

Anderson J Lema

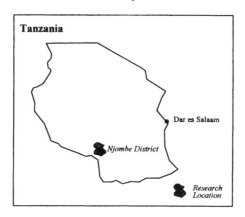

This chapter focuses on an indigenous soil and water conservation (SWC) technique in south-western Tanzania known as *vinyungu* valley-bottom cultivation. In the face of growing population pressures and land degradation, *vinyungu* cultivation, which is largely conducted by women, has come to play an increasingly crucial role in local food security. As an essentially dry-season activity, it is contrasted with upland dry and wet season cultivation with which it is closely linked. However, *vinyungu* cultivation still does not receive the attention it merits owing to the dominant perception among outsiders that it is a side-line, informal agricultural activity.

INTRODUCTION TO THE STUDY AREA AND PEOPLE

Njombe district is a rugged mountainous area in Tanzania's southern highlands.

The altitude ranges from 2000m above sea-level around Njombe town down to 1200m in Makambako division in the north. The rolling topography of the district gives rise to numerous valleys and mountain streams where cultivation is carried out. At lower altitudes the number of valleys diminishes and agriculture is carried out on flatter land.

The main tribal group inhabiting Njombe district is the Wabena tribe. (For historical background, see Culwick and Culwick, 1935; Mwenda, 1963; Graham, 1979; Wardell, 1991.) The Wabena ('people who cut millet') number approximately 315,000 (Bureau of Statistics, 1988). Male labour out-migration is an important feature of Njombe district as it has had a negative impact on the district's economy and household labour supply. Until recently, this has meant that labour for household subsistence requirements has been provided largely by women and children.

Wabena agriculture has been based traditionally on the cultivation of finger millet, sweet potatoes, Irish potatoes, field beans and bamboo. An important famine reserve crop, *numbu*, is grown on poor soils at higher altitudes. While the relative importance of finger millet seems to have declined, the other crops are still cultivated. More recently, hybrid and composite varieties of maize have become the most important crops in the area and are often grown in association with wattle (see photo 19). Throughout much of Njombe district, agricultural crop production is still based on cultivation with traditional hand tools, such as hoes (both traditional and introduced types), machetes, slashers and axes. But in the flatter areas in Wanging'ombe, Makambako and Mdandu divisions, oxen are employed for cultivating land. In some cases, tractors are also utilized.

The soils found in the district are generally of a very low nutrient status and have a poor ability to store and release artificial fertilizers. Typically, these soils occur in the form of red and yellow clays. In some parts of Njombe district, such as Ludewa and Makete, which have a more humid climate, these leached clays have developed on humic topsoil and have a high organic matter content. Techniques to preserve or increase soil fertility thus represent an important component of local farming.

The climate in the study area varies from a semi-arid, warm and tropical climate in the north and north-west, to a cool and high-altitude tropical climate in the centre and south. Rainfall in the district is unimodal, with a single rainy season running from November to May and a dry season during the rest of the year. The mean annual rainfall in the district ranges from around 900mm in the extreme north-western corner to over 1600mm in the southern and western highlands. Detailed studies carried out in the district (and in the region as a whole) confirm the general view that no significant change in the rainfall regime has occurred over the last half century (URT/EEC, 1987). The two main cropping seasons in the district coincide with the wet (*masika*) and the dry (*kiangazi*) season.

VINYUNGU VALLEY-BOTTOM CULTIVATION

During the dry season, the Wabena cultivate small areas of raised beds, known as *kilimo cha vinyungu* in valley bottoms and along stream-beds. *Vinyungu* is a Swahili version of the Bena words *kinyunga* (singular) or *fiyungu* (plural). To a Bena, a *kinyunga* is simply a ridge, a bench terrace or a raised bed, with several of them forming *vinyungu* (see photo 20). In a more general sense, the Bena use *vinyungu* to mean small valley-bottom gardens. *Vinyungus* are an important source of green leafy vegetables, field beans and green maize, as well as providing an additional source of cash income arising from the sale of cabbages, tomatoes, onions and other introduced vegetables.

Vinyungu cultivation is practised in much of Iringa region, but is most common in Njombe district. During fieldwork, no informant could recall precisely when and where *vinyungu* cultivation began, suggesting that the practice is as old as the tribe itself. However, historical accounts of the Wabena, such as those of Culwick and Culwick (1935) would seem to suggest the early 1890s. In the past, *vinyungu* were small in size. Nowadays, they are larger and more extensive, and one *kinyungu* can be considered a garden in its own right.

In Njombe district, *vinyungu* cultivation areas appear as wetland patches spread over the many highland valleys. During the wet season, these valley-bottom wetlands are usually waterlogged or flooded for some time, hindering land preparation or actual cultivation. Most areas of *vinyungu* cultivation have a stream or a river fed by surface run-off from the surrounding upland areas. Subsurface run-off is discharged into the valley bottomlands in the form of springs which may be perennial or seasonal. Water tables are usually high and drainage can be poor at the height of the wet season.

Usually, the smaller and steeper the valley, the fewer the number of *vinyungu* and the smaller the size of the ridges. Large river valleys like the Ruhudji have more extensive *vinyungu*. Valley bottomlands where *vinyungu* cultivation is practised differ from the surrounding uplands in several respects. Not only do they have higher levels of soil moisture during the dry season and in times of drought, but organic matter and soil nutrients accumulate, rendering the soil heavier and richer in nutrients.

The cultivation of *vinyungu* is common among all social classes in the district, with all rural households cultivating at least two or more *vinyungu*. Some of the richer farmers produce large quantities of horticultural produce, especially tomatoes and cabbages, for urban centres like Dar-es-Salaam.

Methods of construction

To prepare *vinyungu*, farmers first clear the land and collect grass and other types of vegetation, such as shrubs, small trees and maize stalks following the harvest. The vegetation is left to dry and, when cultivation time comes, this vegetation is covered by a thin layer of soil to form ridges. An opening is left on the one side at the front or at the back for the purpose of igniting the vegetation.

When the burning of this material is done, farmers use a hand-hoe to mix the ash with the soil to form a rich bed on which they plant their crops. Any solid plant matter that does not burn is broken down into what is locally known as *kubonda tuta*. One ridge or *tuta* is usually 4–10m wide, 10–30m long and 0.5m high. The beds are separated from one another by a 1.2m ditch.

This construction technique is practised on both valley bottomlands (*vinyungu*) and upland plots. On the bottomlands it is known as *kilimo cha mkuha*, while on the upland plots it is referred to as *kilimo cha suvi* or *ngereka*. The only difference between *mkuha* and *suvi* is the location where they are carried out. In both cases it is important to emphasize that normally farmers do not carry out deep cultivation when they prepare the ridges. *Suvi* is suitable for the cultivation of round potatoes, wheat, finger millet, pumpkins and vegetables on the upland plots. On the valley bottoms, *mkuha* is used to cultivate vegetables, maize, beans and pumpkins.

A slight variant of the cultivation practice described above is one in which the burning of vegetation does not take place. As in the practice of *suvi* or *mkuha*, vegetation is collected, heaped in ridges and covered by soil, but it is not burned. Instead, the vegetation is left to rot and the ridges are then planted with crops. This technique is known locally as *kilimo cha kusopel*. It can be found on both upland and bottomlands, although it is more common on the upland plots. Again, deep cultivation is the exception rather than the norm in this kind of cultivation.

In preparing *vinyungu* in the valley bottomlands, ditches or furrows are prepared to separate the ridges from one another. The orientation of these ditches or furrows and the ridges themselves can differ from one place to another. Moreover, ditches can be orientated in different ways within one side. They may lie in a parallel series along the river channel or be perpendicular to the river channel. The main determinant of their orientation is the nature of the slope and the hydrological (drainage) conditions. Where there is a lot of water in the soil (as is often the case where there are several springs), the ditches and ridges are constructed perpendicular to the channel essentially for the purpose of draining the soil. Where the slope is not quite so flat, the ditches and ridges are generally laid across the contour.

Vinyungu construction indicates farmer rationality in land use, and soil and water conservation. The burning of green vegetation is one way of increasing soil fertility. Unburnt vegetation supplies the soil with humus (organic matter), while the burnt vegetation provides the soil with ash (potash). The construction of drainage ditches enhances the environment for plant growth. The construction and planting of crops on ridges itself allows farmers to deal with the high water-table problem. Ridges also help to conserve soil moisture as the growing season progresses. Soil colour and texture in the valley bottomlands is also an important consideration for farmers and allows them to determine its fertility before planting begins. For example, they often refer to *vinyungu* soil as *mganga mtitu* or *lidope fiviya* (ie, black soil). This type of soil can be one of the following: sandy clay loam, loamy sand or clay loam.

Ownership and access to valleylands

In the study area, the number of ridges (*vinyungu*) per household differs from village to village. Most households own from 10 to 20, although some own as few as 2, while others have as many as 30. These ridges may all be located within the same valley or lie scattered over several valleys, depending on the size of the valley and accessibility. The total area of *vinyungu* land owned by households varies from as low as 0.05–0.5ha.

Vinyungu cultivation is essentially a women's activity, although men participate in certain tasks. In most cases, male labour is confined to land preparation, specifically to preparing the ridges. Since women also work in the upland plots, female labour in Njombe is very crucial in ensuring household food security. Women's labour is spread throughout the year on both the upland and valley bottomlands. Hiring labour for *vinyungu* work is not uncommon. Male labour in particular is hired at the time of land preparation for the preparation of ridges.

Farmers usually cultivate valley bottomlands on their own land. However, it is also common for farmers who own little suitable land or who would like to cultivate more *vinyungu*, to request access to valley bottomlands belonging to neighbours or friends in their villages, or even people outside the village. In such arrangements, no transaction of money is involved. As far as the Wabena are concerned, there is no market for *vinyungu* land. Where there is a large river or stream valley, *vinyungu* land is often jointly owned as a common property resource. For example, in one of the valleys studied in Igagala village, some 20 farmers were involved in *vinyungu* cultivation. This practice is common elsewhere.

Joint-sharing of valley bottomlands in Njombe district became relatively common following collective land ownership during the Ujamaa period. At that time all village land, whether upland or bottomland, was placed under the village government authority which allocated it. With the demise of Ujamaa villages and the rise in land value following the liberalization of the Tanzanian economy, access to and control over *vinyungu* valley bottomlands was subject to reform of the land tenure policy.

Insecure land-use rights threaten people's control over the land on which they have built *vinyungu*. Conflicts over access to *vinyungu* land were traditionally unheard of in the study area. However, in the village of Idunda, farmers indicated that the incidence of conflict over valley bottomlands is increasing. In one sense, this demonstrates the growing importance of *vinyungu* cultivation. There is little doubt that produce from *vinyungu* cultivation is an extremely crucial element of local household food security. Not only does it help to meet direct household consumption needs, but a portion of it is marketed to provide cash income.

In some places, *vinyungu* plots are cultivated twice a year to maximize the use of the land and available water. The first planting is carried out in August/September, during which time maize, beans, peas, potatoes and other vegetables are sown. The second planting occurs in March/April. Beans and potatoes are harvested in December and January. This is followed by the harvesting of maize in April or May. By staggering the harvesting of crops in both the bottomlands and uplands there is food to eat all year round.

Farmers suggested that the increasing importance of *vinyungu* cultivation was due to three factors: the growing population, the declining crop yields on upland plots and the vagaries of the weather. Population growth has meant greater demand for food and hence increased dependence on valley-bottom cultivation. Crop production and yield statistics in the district generally show a declining trend or, at least, a stagnation in the upland areas. Moreover, valley bottomlands benefit from upland soil erosion. Fertility here is therefore comparatively higher and, as yields decline on the uplands, more use is made of the bottomlands. There is a common perception among local farmers that in recent times rainfall conditions have changed, with the rains coming later and stopping earlier. Since the valley bottomlands have a more reliable water supply, farmers increasingly are taking advantage of this.

CONCLUSION

The cultivation of valley bottomlands in Njombe district is one of the oldest indigenous land-use practices in southern Tanzania. *Vinyungu* cultivation increasingly has come to play a key role in meeting local household food security, as well as providing a cash income. Over 90 per cent of households in the district depend on products from *vinyungu* cultivation and the signs are that there is a growing dependence on them. Nevertheless, the positive contributions of *vinyungu* cultivation are threatened by a number of factors. One of these is the degradation of the bottomlands. As a result of unsustainable land-use practices, such as over-cultivation around water sources, water pollution and erosion on the valley sides, hydrological conditions in the valleys are changing. The result is that springs are drying up, stream flows are decreasing and there is increased weed infestation. Farmers also suffer from a lack of extension services. The dominant view among district agricultural staff as well as higher policy-makers seems to be that *vinyungu* cultivation is a side-line, an informal land-use activity which is largely confined to women. This kind of thinking prevents them from according *vinyungu* cultivation the attention it deserves.

18

PIT CULTIVATION IN THE MATENGO HIGHLANDS OF TANZANIA

A E M Temu and S Bisanda

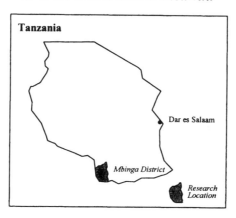

This chapter describes a pitting technique of cultivation known locally as *ngoro* or *ingolu* which was developed by the Matengo ethnic group living in the Mbinga district in the southern highlands of Tanzania. The technique is probably over 200 years old, originating when the Matengo migrated to Mbinga and occupied the forested mountains, where they lived in caves to protect themselves against the warring Ngoni, another ethnic group which moved in from South Africa. Faced with possible starvation, the Matengo had no option but to cultivate crops on the steep hillsides. The pitting technique developed by the Matengo to control soil erosion and to improve soil fertility is still used today and serves as a good example of a successful indigenous technology which controls erosion on steep hillsides.

INTRODUCTION TO THE REGION AND THE PEOPLE

The Mbinga district borders Mozambique to the south and Lake Malawi to the west. Covering a total area of over 11,000km², over a quarter of the region is under water and another quarter is made up of natural forest. Arable land takes up the remaining three-quarters of available land, out of which only about 16 per cent is cultivated.

The Matengo Highlands are located in the central part of the Mbinga district, extending from an altitude of 900m along Lake Malawi to over 2000m above sea-level. The climate in the highlands is temperate to cool tropical with maximum temperatures reaching 30°C and a minimum of 13°C in the higher altitudes where there are also occasional frosts. The onset of the rains is fairly reliable, starting in October or November, with the main wet season between December and April. Total rainfall is in the range 900–1200mm.

The soils in this area are deeply weathered, reddish or brownish sandy clay, with variable organic carbon and pH levels (oxisols and ultisols). The depth of soil varies with elevation; at high altitudes the soils are less than 2m deep with a brown, sandy clay topsoil. At lower elevations of less than 1500m, the soils are very deep (more than 3.5m), well-drained red sandy clay soils.

Ethnically, the Mbinga district has four major groups, all of whom have a distinct dialect and a different culture. The Matengo, whose name comes from the word *kitengo* which means woodland (forest), referring to people who live in forests, are the biggest ethnic group in the area. They are now found mainly in the central highlands, the most densely populated part of the Mbinga district, while the Nyasa, Manda and Ngoni ethnic groups live in the less populated eastern coastal strip of Lake Malawi and the eastern part of the district. The average population density of the region is low, estimated to be 35 people per km², although areas around Mbinga town in the central part of the district have up to 120 people per km².

Using and owning land

Land is the most important resource to farming families living in the Matengo Highlands. Ownership is communal with individual user rights. Land allocation is provided free of charge by the village council. Once land is given to the household head (usually a man), the household has the right to use that land for life. No one is allowed to sell land, although some families may sell it secretly.

Within the household, land is passed from the father to the elder son. Women are not able to inherit land, but they obtain user rights from their husbands or fathers. Land borrowing is practised, although it is not common in high-density areas. In most cases, presents are given as a sign of appreciation to the lender. An important aspect of borrowed land is that the borrower is not allowed to grow permanent (perennial) crops on their land. Grazing fields are communally owned in less populated areas. In the Matengo Highlands, all animals are tethered because there is not enough land available to graze livestock. Woodlands are also communally owned, although individual households have their own woodlots for shading coffee as well as firewood. The average farm size in the Mbinga district is around 1.5ha, although this may vary in the Matengo Highlands region where land is more scarce.

There is currently a wave of migration away from the central Matengo Highlands to less populated areas in the district. In some cases it seems that parents encourage their sons to go and live in areas where land is not scarce, for example to the eastern part of the district. As land is more abundant, fallows are common and last longer, up to seven years. Also, cultivation up to the top of the hills is not necessary in areas with plenty of land. Most importantly, migrants leaving the highlands to less densely populated regions continue to use the pitting system wherever they settle, thus spreading the practice beyond the Matengo Highlands.

Cultivating the highlands

A wide range of crops is grown in the Matengo Highlands. The main staple cereal crops are maize, wheat and finger millet. The major pulses grown are beans and peas, while root crops include cassava, sweet potatoes, round potatoes and cocoyams. Sunflowers are also grown as an oil crop. Vegetables are grown in the valley bottoms throughout the year and fruits are also grown. Bananas are cultivated for food (green bananas) or are allowed to ripen and are eaten as a fruit. Coffee (arabica) is the major crop grown in the highlands. With the exception of coffee, fruit trees, bananas and vegetables, all other food crops are grown under the pitting system.

THE PITTING SYSTEM

History

The first settlements in the Matengo Highlands were around Litembo, to the west of the present Mbinga town. Litembo is said to be 'the centre of diversity of the pitting system'. Faced with the need for survival, the Matengo people had no option but to cultivate the steep hillsides. When the elders of over 65 years of age were asked when the practice was developed, they could only recall that their great grandparents had constructed pits. The first Germans reaching the Matengo region in about 1890 thus found the Matengo tribe in the fertile Litembo area practising the pitting system.

The role of pits is to control run-off water from eroding soil on the steep hillsides. The rainwater is collected in the pits, percolating into the soil without causing any harm. If a pit is not deep enough, the water overflows into the next pit on the slope, thus reducing the amount of water and soil lost by surface run-off (see photo 21).

Preparation

Three stages are involved in the preparation of the pits:

1. The grass is cut in the field where the pits will be dug.
2. This is collected when dry and laid out in strips, forming grids all over the area earmarked for cultivation. These may be square or rectangular, with sides varying from 100 to 200cm.
3. The grass is then covered with topsoil, which is dug from the centre of the grid and spread on top of the grass leaving pits (*ingolu*) in the middle. The pits vary in depth from 15 to 60cm, and from a distance resemble a honeycomb or chessboard. This digging work is done exclusively by women.

Men use a simple tool called a *nyengo* to cut the dry grass. The tool resembles a manual grass-cutter (slasher) with a curved end. The arrangement of grass into grids is done manually and, with experience, the size of the grids is estimated. Women use a simple hand-hoe to scoop the soil from the middle of the pit, making sure that the grass is fully covered with the topsoil before they move on to the next pit. The ridges surrounding the pits are strengthened by the grass, thus minimizing the chance of breakage during heavy storms. Landslides, which could be potentially disastrous, were not apparent, possibly because the soils generally have a high infiltration capacity.

Although the pits are generally effective in preventing soil erosion (as evidenced by the low level of sedimentation in the rivers and streams during the rainy season), they are highly labour intensive compared to flat land cultivation. In spite of this, the pitting system is used by over 70 per cent of the farmers in the Matengo Highlands. Most of the pits are found on slopes of around 50 per cent, the minimum slope being 10 per cent, while the steepest slope was over 60 per cent.

Maintaining soil fertility

Pits that are dug when the land is first cleared, or after a long fallow period, are cropped with finger millet, beans or cassava. In the south of the district where forest is being cleared, all farmers plant finger millet as the first crop (ICRA, 1991). In the Hagati Plateau, where the fields are cultivated intensively, most of the farmers plant cassava after the fallow. However, during annual rotations, over two-thirds of the farmers plant beans as the first crop. When asked why they prefer beans, farmers answered that 'it has the fertilizer effect'. Farmers sow the seeds on the ridges which are underlain with compost and then covered by subsoil dug from the centre of the pit. During weeding, all weeds are thrown into the pit, where they are left to form a compost. At the end of the season, all crop residues are incorporated into the pits. Beans are usually planted in March/April and harvested in June. If the pits are planted initially with wheat or peas, they are also harvested in June. In the following season, the same pits are rotated with maize planted in November/December. When maize is harvested, the ridges are split: new pits are formed in place of the former ridges, and new ridges in place of the old pits. Thus the cycle of rotation starts again.

Length of fallow and soil fertility

The length of the fallow period differs between regions in the Matengo Highlands. Where population density is high, the mean number of years of continuous cultivation on the same field is 6.5 years (ICRA, 1991). In the less populated areas, this is reduced to 4 years. Fallowing is very common in the highlands and the length of fallow depends on the amount of land a farmer owns and population pressure. Consequently, the length of fallow period varies greatly, from 2 to 10 years, although the most common length of time is 3 years of fallow. Because of the short periods of fallow and longer periods of continuous cultivation, soil fertility appears to be decreasing. A farmer of over 70 years of age said that a long time ago they used to harvest around 3.75t/ha, or, in his terms, 15 bags per acre. Today, they only harvest around 7 or 8 bags per acre. It is important none the less to realize that if it were not for the incorporation of organic matter in the pitting system, the rotation of crops and some fallowing, yields would be even lower.

Land use patterns

The factors determining land-use patterns around settlements are local knowledge of soil types and labour availability. For instance, coffee and banana, both labour-intensive crops, are normally planted close to the house. This reduces the amount of time needed for collecting water to spray fungicides and for pulping in the case of coffee. Immediately adjacent to coffee come staple crops, such as maize, beans, potatoes, peas, wheat, cassava and finger millet, depending on the soil type. For instance, finger millet and sweet potatoes are grown on newly opened land or immediately after the fallow. Valley bottoms are usually planted with early beans planted in November. Fruit trees are also planted near the homesteads for ease of harvest.

Labour

Traditionally, the division of labour is largely determined by gender. At the beginning of the season, men clear the fields and arrange the dry grass in the form of rectangles. Then women (adults and young girls) dig the pits during March/April. This is the most labour-intensive and laborious task. A strong woman can dig up to 150 pits in a day and will take up to 6 days to complete 1 acre (< 0.5ha). In the flatter areas bordering the highlands where ridges are used, men, women and children participate in their construction. However, preparing the pits is strictly a woman's task, whereas planting and weeding are shared activities. Harvesting is likewise a shared activity, while crop processing is predominantly the responsibility of women in the household. Women are therefore essential to the pitting system used by the Matengo.

Generally, labour is provided by family members, although during the peak period some households run short of labour and therefore either hire paid labour or they organize labour parties known locally as *ngokela*. The total labour requirement for 1ha is about 15 person days to make around 2250 pits. Labour-saving technology, such as the use of tractors or oxen, is non-existent in the Matengo Highlands because of the gradient of the hills.

CONCLUSION

Pike (1939) reported that 'since the advent of the British rule, many farmers moved to more fertile parts of the highlands and that the further they moved away from Litembo, the less inclined are they to follow the system which the tribe evolved for its safety in the past'. Stenhouse (1944) later added that 'the younger Matengo are adopting lazier methods of cultivation, while the introduction of cash crops is accelerating this deplorable breakdown'.

Contrary to these beliefs that the Matengo have been abandoning their pitting system for 'lazier' types of farming, this study did not see any indications that this was occurring. Farmers repeatedly declared that as long as they live on the steep slopes of the Matengo Highlands, they will continue to make pits and grow crops on them. They did not want to risk losing their good soil to the valley bottoms through erosion, and all farmers recognized the significant role that the pitting system plays in maintaining and improving soil fertility. The pitting system is expanding rather than contracting, extending with migrating farmers into new areas of Mbinga.

The government of Tanzania fully supports the pitting system and has often funded farmers from elsewhere in the country to visit the Matengo Highlands to study the system and adopt it in their area. Soil erosion is a serious problem in many parts of Tanzania, largely through heavy deforestation, overgrazing, uncontrolled fires and farming systems which do not incorporate techniques to conserve soil and water. Using the pitting system, farmers on steep hillsides which are even more prone to soil erosion have been able to sustain modest yields from their plots for several decades with minimal or no mineral fertilization. New interventions to improve the pitting system will ensure the continued use of this indigenous system and perhaps its expansion into less hilly areas elsewhere in Tanzania and beyond.

19

'BACK TO THE GRASS STRIPS'

A history of soil conservation policies in Swaziland

M Osunade and C Reij

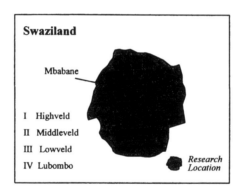

The evolution of soil-conservation policy in Swaziland has been influenced greatly by the views of experts regarding the severity of erosion. These views, however, have been based on subjective assessments with little or no hard data to support the view taken. No effort has ever been made to analyse and measure erosion processes with any precision in Swaziland, which is quite surprising given the large amount of money and energy invested in soil and water conservation over the years. This chapter outlines the varied policies and practices adopted in Swaziland to combat soil erosion in recent decades, and describes the strong attachment by farmers to grass strips as a means of reducing run-off.

INTRODUCTION TO THE COUNTRY

The kingdom of Swaziland is one of the smallest African states, with a size of 17,346km². Although the country is small, it has four major geographical zones which run from north to south: the Highveld, Middleveld, Lowveld and Lebombo. Rainfall decreases sharply from Highveld (1500mm) to Middleveld (900mm) and Lowveld (< 550mm) and Lebombo (> 550mm). Most Swazi live in the first two of these zones, where average population densities are over 50 persons/km^2. However, owing to the growing land shortage in both regions, people are now moving increasingly into the lower lying areas where rainfall is scarcer.

Swaziland has two major land tenure systems: Swazi Nation Land and Individual Tenure Farms. In the former, land is held in trust by the king for the Swazi people, but is administered by the chiefs. The average size of the land cultivated per homestead is less than 2ha and the dominant crop is maize. By contrast, Individual Tenure Farms are usually large commercial farms. Although the population is heavily involved in cultivation, Swaziland is traditionally a pastoral nation. The number of cattle was estimated to be 740,000 in 1991, which means on average about one head of cattle per inhabitant.

THE HISTORY AND EVOLUTION OF SOIL CONSERVATION IN SWAZILAND

The pre-1949 period

In the annual report of the Veterinary and Agricultural Department for the years 1942–46, mention is made of some efforts in the field of soil conservation:

- Contour lines were laid out on native and European lands, which were planted with napier grass.
- Eroded areas were fenced off for demonstration purposes and showed rapid recovery which, according to the Veterinary and Agriculture Department Report (1946, p 29) indicated that the solution to erosion had been found.

The period 1949–1960

During the 1940s the colonial agricultural officers believed that the country's natural resource base was degrading rapidly owing to the very unequal distribution of land between the Swazi and the European populations, and overgrazing of rangeland areas (Faulkner, 1944; Scott, 1951). As a result, as part of the 1948 Eight Year Development Plan, a major soil conservation programme was launched in 1949 with its main focus on introducing grass strips.

During this period, more than 70,000 miles of grass strips (almost 114,000km) were laid out (see photo 22). It was assumed that each mile of grass strip protected about 5 acres of land, which meant that in 1960 more than 350,000 acres (140,000ha) had been protected against soil erosion. This was only achieved because the late King Sobhuza II gave his full support to the conservation of Swaziland's natural resources. Without his explicit backing, grass strips would never have been established on this scale in such a short period.

The period 1960–1970

In 1960, the total length of grass strips laid out fell, for the first time since 1949, to under 1000 miles. This was mainly because almost all cultivated land had grass strips by this time. In 1966, the government of Swaziland declared that 'the immediate and prime goal of Government agricultural policy is the rapid development of agriculture in the Swazi Nation Areas'. This would be achieved by the selection of a number of Rural Development Areas (RDA) in which co-ordinated development programmes were to be undertaken. Soil conservation activities were part of the RDA programme because it was the government's opinion that soil erosion measures were not keeping pace with erosion. USAID decided to fund soil conservation activities in the first four RDAs, but its consultants decided to move away from grass strips because these were not believed to be efficient. They even proposed to remove grass strips and to replace them by a system of terraces, despite the lack of evidence for the weaknesses attributed to the former.

The period 1970–1977

By the 1970s, in a seminar on 'Overgrazing and livestock development' (Ministry of Agriculture, 1978) the situation was described in the following way:

> In many areas all vegetative cover has been lost and the soil is exposed to erosive forces. Immediate measures are needed on 228,000ha ... at the present rate of soil loss over 30,000ha have a life expectancy of less than 50 years and thereafter will be barren.

A report by USAID (1978: 16) described the situation in equally dramatic terms for cultivated land: '... much of the surface soil has been lost ... and the remaining root zone of much of the crop land is less than 45cm in depth'.

A major terracing programme was developed to address the apparent urgency of soil erosion. Heavy earth-moving equipment for terrace construction was bought by the government of Swaziland with a loan from USAID, which also provided a grant for technical services, training and other aid assistance. From 1971 to 1977, about 7200ha were terraced and in the process grass strips were removed. At the end of this period, strong criticism of terracing emerged: 'in the past few years, the standard of soil conservation measures, viz terracing and grassed waterways, have proved to be expensive, land consuming and some-times even erosion inducing' (Spaargaren, 1977). Instead, Spaargaren recommended the use of properly designed, maintained and propagated grass strips. His recommendations were adopted, to a large extent, by USAID's feasibility study team which in 1977 examined the cost-effectiveness of soil conservation practices.

The period 1977–1983

Despite the fact that terracing was severely criticized in 1977, this was not reflected in the targets for the second phase of the RDA programme. The workplans for the 18 RDAs indicated the following targets: terracing of 16,500ha, removal of grass strips on 17,950ha and removal and realignment of grass strips on another 5500ha. USAID provided another loan for equipment as well as a grant for technical assistance. In reality, the terracing programme fell far short of its targets and few grass strips were removed.

The period 1983 to date

The RDA programme ended formally in 1982 and activities could not be continued on the basis of government funding alone. No major conservation works were carried out in this period because soil erosion was no longer considered to be a threat, and hence the priority of the government shifted to other problem areas. Recently, the interest in soil conservation seems to be increasing again. In 1988 a study was undertaken on gully erosion in Swaziland (WMS Associates, 1988) which showed that, in particular, the Highveld and Middleveld have many active gullies owing to a combination of overgrazing and cattle tracks. The major drought which hit Southern Africa in 1992 also affected Swaziland and livestock mortality was high. As a result, government and land users have become more aware of the limits to their natural resources and several NGOs have now initiated SWC activities.

The importance of grass strips

The present soil conservation wisdom asserts the need to move away from conservation structures to low-cost vegetation barriers. In this respect, the Swazi did 40 years ago what is now recommended by soil conservationists. Grass strips laid out in the 1950s remain a common feature, in particular in the Highveld and Middleveld, on small Swazi farms, although not on large-scale commercial farms. The grass strips were introduced systematically because King Sobhuza II issued an Order-of-the-King in 1954 obliging all Swazi to install grass strips on their land. Although King Sobhuza II died 15 years ago, the Swazi farmers generally continue to maintain the grass strips on their fields. They may not always follow the contour, and ploughing may have reduced their width, but the accumulation of sediment behind the grass strips has led gradually to the formation of terraces in many places.

Swazi farmers perceive a number of advantages and disadvantages of grass strips. Most farmers in the Highveld, Middleveld and Lebombo, where there is little flat land to cultivate, state that it is impossible to farm without grass strips, and they install them on all new fields. Grass strips are also important to women, because they are a source of raw material for their handicraft activities, and during the winter they provide fodder. The disadvantages mentioned by the farmers are that grass strips reduce the area that can be cultivated, provide a home to rodents, and slow the speed at which land can be ploughed by a tractor. Although there are cases where grass strips have been abandoned, the general trend is towards maintenance and expansion, suggesting that for farmers the advantages outweigh the costs.

CONCLUSION

Land pressure continues to build up in Swaziland and it is important to investigate whether productivity can be increased. Grass strips are created by leaving land unploughed, the strips being composed of grasses which appear spontaneously. *Hyparrhenia* sp, a grass used for thatching, is one of the most common grasses. No efforts have been made to introduce new grass species or shrubs, for instance, to increase the fodder value of the grass strips, although improving the quality of the grass strips would face other constraints. First of all, after the harvest, cattle graze on the stubble so, unless grazing is controlled, it would be difficult to introduce fodder grasses. Vetiver grass could improve the efficiency of the present grass strips in terms of a further reduction of soil loss, but its low fodder value makes it unlikely that Swazi farmers will be willing to invest time and effort in planting strips of this grass. Secondly, it is important to bear in mind that women have a specific interest in grass strips for their handicraft activities and any effort to change the quality of grass strips should take their needs into account.

Thus, grass strips would seem to have an important role in future soil conservation practice if they can be adapted to serve the diverse needs of Swazi farmers.

20

THE 'FLEXIBILITY' OF INDIGENOUS SOIL AND WATER CONSERVATION TECHNIQUES

A case study of the Harerge Highlands, Ethiopia

Kebede Asrat, Kederalah Idris and Mesfin Semegn

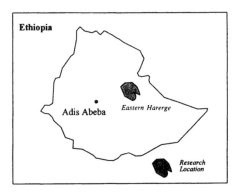

In the face of growing population pressures and a decline in cultivable area, farmers of the Harerge Highlands in Ethiopia have become increasingly reliant on a unique set of soil and water conservation (SWC) techniques. These techniques enable them to exploit local resources sustainably and to survive in their rugged terrain. However, owing to perceptions that the area is exposed to a high risk of land degradation, the Harerge Highlands have been targeted by outside donor and government agencies for assistance with large-scale SWC programmes. As we argue in this paper, project interventions often fail to address the diversity of local circumstances and needs, and can undermine the flexibility of traditional approaches to SWC.

INTRODUCTION TO THE STUDY AREA AND PEOPLE

The Harerge Highlands cover an area of approximately 1,500,000ha, stretching from the Gelemso-Asebeteferi chain of mountains in the west to the Kundudo Mountains in the east. To the south lie the Ogaden Deserts and to the north the Danakil or Afar Desert. The Harerge Highlands consist of hilly terrain of low to moderate relief, interspersed throughout with sloping valleys. The high population density, combined with the rugged topography and the erosive tropical storms which are common in the area, make it a challenging place to be a farmer.

This study was conducted in three woredas in the East Harerge administrative zone. A woreda is the smallest administrative unit in Ethiopia and is composed of a number of Peasant Associations. The woredas were selected to reflect the basic features of the three agro-ecological (altitudinal) zones and the diverse farming systems found in the Harerge highlands (see the table below). Only 10 per cent of the land area of the woredas lies in the high altitude *Dega* agro-ecological zone, while the *Woindega* and *Kolla* zones make up 71 and 19 per cent respectively of the total area.

The distribution of agro-ecological zones by woreda

Woreda	Area of agro-ecological/altitudinal zones (ha)		
	Dega (high)	*Woindega* (medium)	*Kolla* (low)
Fedis		33,359	1,756
Alemaya		39,840	7,031
Meta	13,857	24,943	16,629
Total	13,857	98,142	25,416

The study woredas are experiencing tremendous population pressures. The total population is approximately 407,400 people (Ministry of Agriculture/Shawel, 1988), with an average family size of 6.4. The population density ranges from 231 persons per km^2 in Fedis (in the lowlands) to 281 and 409 persons per km^2 in Alemaya and Meta respectively. Age-structures show a high proportion of children. While polygamy is common and often practised by relatively better-off farmers, some 5 per cent of the households are female-headed. The majority of the farming population belongs to the Oromo ethnic group, although some Amharas can be found in the highlands of Meta.

The following table illustrates the different patterns of land use in the study woredas. Considerable differences are also apparent between the altitudinal zones. In the higher altitude *Woindega* and *Dega* zones, much of the terrain is too sloping for cultivation. Owing to the relatively higher population pressure, there is a shortage of land for cultivation and livestock raising. In the lowlands (*Kolla* zone), land availability is not such a problem. The proportion of land for grazing and cultivation tends to be higher here.

Most of the soils in the study woredas are loamy with a few clay-textured

Land use patterns by woreda

	Fedis	Woredas Alemaya	Meta
Household numbers	16,544	25,064	22,066
Total cultivated land (ha)	35,722	24,687	14,890
Average cultivated land per household (ha)	2.16	0.98	0.68
Total grazing land (ha)	3,827	1,077	520
Average grazing land per household (ha)	0.23	0.04	0.02

soils. Their colour ranges from dark grey to dark brown, with some reddish and black soils. The soil is often stony and in some areas rock outcrops are common. The pattern of rainfall in the study area is bi-modal and often erratic, with average rainfall ranging from 700 to 1100mm annually. The main rainy season (*kiremt*) lasts from June to September, while the shorter rainy season is between February and May. Local farmers report that in recent years the pattern of rainfall has become more unreliable, making it difficult to determine the best time for planting.

The dynamics of change in local farming systems

Farmers have responded in various ways to the changes that have occurred in recent years in the Harerge Highlands. Owing to increased demographic pressures and the frequent redistribution of land by Peasant Association executive committees, land holdings in the region have become greatly fragmented. Farmers can acquire land through different means, including inheritance, share-cropping and the clearing of uncultivated land. Farmers have use rights over the land they own but, according to the policy of the transitional government, cannot sell it. Of the 150 farmers in this study, only 24 per cent had a single plot, while the remaining had two or three plots.

The expansion of agricultural areas through the clearing of bush seems to have reached its peak at present, especially in the highland areas. In recent history, this expansion has been most marked when changes of government have occurred. This was the case in 1974 when the feudal monarchy was overthrown and again in 1991 when the socialist regime of Mengistu fell. Large forested areas which had been either closed off or under community afforestation programmes were cleared by farmers.

Growing pressure on both farming and grazing lands has increased the importance of mixed farming whereby there is a close interaction between animal husbandry and crop production. Maize and sorghum thinning, as well as crop by-products (including tree leaves), are used as a source of feed for livestock. Cattle in turn provide draught power and manure (*dhikke*) which is used for fertilizing the crops. The use of inorganic fertilizers is a more recent and growing phenomenon (in the last 30 years), spurred by the decline in soil fertility and the expansion of cash crop production.

Livestock thus remain an integral component of the farming system, although in recent years the number of livestock per family has decreased in all zones. This is attributable to a shortage in fodder (grazing land) caused by the expansion of farming, although cattle diseases and the sale of livestock to meet pressing household needs are also factors. Farmers tend to prefer goats to sheep and cows over oxen, with over a half of families owning one or no oxen. Cows are preferred, not only because they provide milk and other marketable products, but also because they produce calves or heifers for sale and can be exchanged under cattle share-keeping agreements. Farmers manage ploughing instead with *dongora* and *akaffa*, which are local agricultural tools.

While coffee and the stimulant, chat, have traditionally been the most common perennial cash crops in the study area, the expansion of the access road network together with increased marketing opportunities have made chat the predominant crop. The opening of new markets in neighbouring countries has also resulted in the cultivation of potatoes and the introduction of other high-earning vegetable crops, such as cabbage, onions, egg plant, etc.

Agricultural productivity in the study area is thus increasingly constrained by the decline and unreliability of rainfall, the decline in soil fertility (including the over-utilization of land), the shortage of fertilizers (both traditional *dhikke* and artificial ones), the shortage of plough oxen, and the prevalence of pests and diseases. The problems of soil erosion and degradation in particular increase with altitude. These difficulties have contributed to an out-migration of farm labour to urban centres in certain areas. In others, the changing dynamics of local farming have created new opportunities which have led to the emergence of a new class of farmer-traders.

TRADITIONAL SOIL AND WATER CONSERVATION TECHNIQUES

As agricultural productivity has fallen owing to changes in the local environment, traditional SWC techniques have increased in importance. These techniques cover a wide range of agronomic, biological and mechanical measures which draw on accumulated knowledge and which have evolved incrementally in response to local needs. They include crop rotation, intercropping and the planting of cactus or strips of perennial grass which serve to diversify production, increase soil fertility and reduce erosion. The construction of stone and soil bunds, in particular, illustrates how the flexibility of local techniques has been threatened by the introduction of structures from outside.

Stone bunds

Stone bunds or stone lines (known locally as *dhagga*) are barriers of stones placed at regular intervals along the contours of farms (see photo 23). This technique has been used in the study area for generations and is most commonly found in the *Woindega* and the *Kola* agro-ecological zones. Some sources attribute the introduction of stone bunds and other local ethno-engineering practices to traders coming from Saudi Arabia, where stone-terrace technology for the cultivation of chat (*Catha edulis*) and coffee was employed (Ministry of Agriculture/Shawel, 1988).

The size of stone bunds varies, depending on the availability of stone and the topography. In sloping areas and where stone is readily available, bunds are closer together and relatively higher and wider than in areas which have a gentler slope and where stones are scarce. Stone bunds are generally constructed incrementally and extended from year to year as needed. While the average height of stone bunds in the fields was approximately 0.5m, with slightly smaller dimensions for the width, it is not uncommon to find older stone bunds measuring 2m high.

Farmers construct stone bunds for different reasons. In areas where fields are stony, farmers clear the field of stones and lay them along the contour so that they can plough. In other cases, bunds are deliberately constructed for soil and water conservation purposes. Stone bunds are also used to retain or slow down run-off and to prevent erosion. Their impact on crop yields owing to the increased moisture infiltration and the decrease in nutrient loss is particularly pronounced in the relatively drier areas.

Soil bunds

Soil bunds are found most often in *Woindega* and *Kolla* zones, especially in fields where chat, maize and sorghum are cultivated. Like stone bunds, the methods of construction and the spacing and dimensions vary, depending on the purpose and the area. However, unlike stone bunds, they can be destroyed more easily and rebuilt as needed.

Bunds in fields cropped with maize or sorghum are constructed at the time of land preparation or after planting. The bunds are constructed by digging a trench about 25cm deep and by forming an embankment on the lower side of the trench. After planting, when erosion occurs, usually following a large storm, bunds are constructed by digging the soil and forming embankments or ridges. Here, the *akaafa*, the traditional inverted flat hoe, is used. These bunds are often destroyed after a period of six weeks to control the spread of a dangerous weed known as *buranna* (*Digitaria* sp).

Soil or stone-faced bunds are built in fields where chat is cultivated. They are designed for both soil and water (either rainfall or irrigation water) conservation and are usually constructed at the time of land preparation. In some instances the bund is constructed one year after planting of the chat crop in order to ensure that the field is free of buranna weeds. However, unlike the soil bunds on cropped fields, bunds on chat fields are permanent and last for the entire rotation of the crop, which is usually 20–30 years.

The spacing of the chat bunds observed ranges from 2 to 5m on steeper slopes and from 5 to 10m in the gentler sloping areas. In fields which have moderate to deep soil, gentler slopes, and a lower risk of erosion, the bunds are made of soil and are relatively large. In areas with shallow soil and relatively steeper slopes, the bunds are lower and, if stone is available, are usually stone-faced.

Hand-built modified bench terraces are the result of the benching effect of bunds and unploughed strips of land. Farmers confirmed that many years ago strips of unploughed land were left fallow. With time, the strips trapped sediment, which grew in height until eventually they formed terraces. These terraces are usually used for the cultivation of chat, coffee, fruit trees and vegetables, and are most often found in the valleys. The terrace risers are usually stone-faced and slightly slanted inwards, and can attain a height of 1m or so. On steeper slopes, the terraces are narrower and the rise is higher.

Introduced bunds

Owing to the perception that land degradation is a serious problem, the Harerge Highlands have been targeted by large-scale SWC programmes promoted by organizations such as the World Food Programme, the Food and Agriculture Organisation and the Ministry of Agriculture in Ethiopia. There is a significant difference between traditional and project-introduced methods of constructing stone bunds and other ethno-engineering devices, which has a bearing on their adaptability to local conditions.

Traditional methods usually involve an incremental and adaptive approach which responds to the dynamics of changing conditions. For instance, depending upon the availability of labour and stones, construction is carried out in phases, starting with a line of stones only a few metres long, which is gradually extended over time, usually a period of years, until the entire field is covered. The spacing and size of the structures are determined and adjusted by the farmer after seeing the pattern of run-off flows in times of rainfall. By contrast, the project method usually emphasizes a rigid, one-off approach. Structures are predetermined, with little attention being paid to the physical characteristics of the cultivation area. The structures tend to be based on simplistic engineering formulations which are slope-based.

Introduced methods do not draw adequately on the accumulated and specific knowledge which farmers have. Construction work is generally carried out without consulting the farmer who owns the land about the spacing or the dimensions of the structures. Local techniques, on the other hand, evolve as a part of farming performance, which is a product of local circumstances. One example which local farmers gave was that in the traditional method of construction, consideration is usually given to passages and steps for livestock and humans. In the project method, this is generally disregarded. This contributes to the collapse of bunds on crop fields.

Although introduced methods can restrict flexibility and adaptability, they also offer useful lessons. Some farmers in the study area noted that their knowledge of how to incline risers was enhanced by observing the construction of introduced bunds. While this may be the case, more care needs to be taken when large-scale SWC techniques are brought in by outsiders. For instance, where these techniques are introduced through development or food-aid subsidies, they may disrupt traditional labour markets in ways which are not yet fully understood.

CONCLUSION

Farmers in the Harerge Highlands of Ethiopia have become increasingly reliant on a diverse set of SWC techniques owing to growing population pressures and the associated decline in cultivable land. This includes a variety of traditional mechanical, biological and agronomic techniques which have been adapted to the different crops, labour demand patterns and physical conditions found in the three agro-ecological/altitudinal zones. As the case of soil and stone bunds illustrates, structures are built incrementally responding to local needs and circumstances. Introduced structures fail to do this. They have been imposed, based on simplistic engineering formulations, and in the long run they may disrupt the flexibility upon which farming performance in traditional systems is based.

17. *Semi-circular bunds support a growing stand of maize, Zimbabwe.*

18. *Raised bed cultivation of maize, northern lakes depression, Zambia.*

19. *Wattle and maize in association, Itulike village, Tanzania.*

20. Vinyungu *farming in Itulike village, Tanzania.*

21. Matengo *pits in Mbinga district, Tanzania.*

22. *A maize crop growing between grass strips, Swaziland.*

23. *Stone bunds in Harerghe region, Ethiopia.*

24. *Traditional ditches on farmland, northern Shewa, Ethiopia.*

25. *Drainage ditches in northern Shewa, Ethiopia.*

26. *View of a hillside treated by conservation measures, near Konso, southern Ethiopia.*

27. *Combining conservation measures: permanent stone bunds, terraces and strip cropping, central Ethiopia.*

28. *Traditional hedged hamlet in the Transkei, South Africa.*

29. *Fence of straw and brushwood, Transkei, South Africa.*

30. *Erosion of contour bunds, Transkei, South Africa.*

31. *Vertical ridges in a field of manioc, maize and young banana plants. In the foreground, a hay mulch protects land still to be ridged. Bakou, Cameroun.*

32. *Intensive use of land in Bafou, Cameroun. Trees and hedges supplement contour ridges to conserve soils.*

21

TRADITIONAL DITCHES IN NORTHERN SHEWA, THE ETHIOPIAN HIGHLANDS

Million Alemayehu

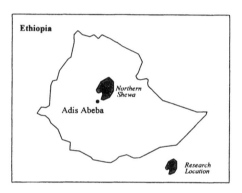

In many parts of the Ethiopian Highlands, land degradation owing to soil erosion has become a serious problem. For more than a decade the government of Ethiopia has undertaken a massive programme of soil and water conservation works, particularly the construction of terraces. However, in most highland parts of Ethiopia, particularly in northern Shewa, it has been observed that some of these terraces have been destroyed by the local farmers, while certain indigenous soil and water conservation methods such as stone terraces, cut-off drains, etc, continue to be used. Among these indigenous soil and water conservation techniques, traditional ditches have been used widely by farmers for different purposes in many parts of the highlands (see photo 24). Traditional ditches are constructed every cropping season and run diagonally over the cultivated land. They are made by pressing a *maresha* plough deep into the ground and can be differentiated easily from the normal plough furrows.

This chapter looks at the factors which influence farmers' use and assessment of traditional ditches. Participatory Rural Appraisal (PRA) techniques such as transect walks were used in this study, as well as formal individual interviews using a questionnaire. Informal interviews and discussions were conducted with individuals as well as groups in order to obtain more information on the perceptions of local farmers.

INTRODUCTION TO THE AREA AND THE PEOPLE

From Keyit Wereda in Northern Shewa, two Peasant Associations, Andit Tid PA and Denqwan-Gunagunit PA were chosen for this study as they are typical of areas where traditional ditches have been used widely under many different conditions.

In both areas the average annual rainfall, which falls in two rainy seasons, is about 1350mm. The main rainy season (*meher*) normally lasts from June to October and the short rainy season (*belg*) lasts from March to May. Farmers in both areas are involved in subsistence crop production under rain-fed conditions with livestock husbandry. The major crops grown are barley, beans, peas, wheat, lentils and linseed. In general, cereals are used for consumption, while pulses and oil seeds are used as cash crops. Perennial crops are not grown, probably because of the cold climate at this altitude. Both areas are high with steep cultivated slopes, ranging from 2850m above sea-level in Denqwan-Gunagunit to between 3000–3500m above sea-level in Andit Tid. The soils of Andit Tid are andosols, regosols (known locally as Tikur Afer and Merere respectively) and cambisols, with texture varying from medium to coarse (Bono and Seiler, 1984; Yohannes, 1989). Although no detailed soil survey has been carried out, the soils of the Denqwan-Gunagunit are mostly dark brown to dark clays, locally known as *tikur merere*, varying in texture from heavy clay to sandy clay loam. The fertility of the soil in both areas varies from low to relatively high.

There are no recent shifts in crop cultivation, although in Andit Tid there are some eucalyptus plantations on communal land. Most of the farmers have a negative attitude towards these plantations as they do not benefit from them in any way. In both areas many farmers have been planting eucalyptus around their homesteads and on the margins of their cultivated land, as it is fast growing and produces firewood and construction materials within a short period of time.

Livestock production is an essential part of the farming system as nearly all land preparation is done with ox-drawn ploughs. Sales of animals and animal produce contribute significantly to farmers' income. Horses and donkeys are used for transport, while small ruminants and poultry provide food and a convenient source of cash.

Although both areas are almost the same in terms of rainfall and crop patterns, there are some important differences between them. Andit Tid PA, with an estimated population density of 69 persons per km^2, is less densely populated than Denqwan-Gunagunit PA, which has an average density of 205 people per km^2. Andit Tid PA covers an area of 1840ha, whereas Denqwan-Gunagunit PA covers only 720ha. The average household size of 3.7 persons in Andit Tid PA is also considerably smaller compared to 6 persons in Denqwan-Gunagunit PA.

In Andit Tid PA, new soil and water conservation measures (SWC) (*fanya-juu*, soil bunds, waterways, etc) were introduced in 1982 by the Ministry of Agriculture which are still visible on some of the cultivated plots of land. In Denqwin-Gunagunit PA, the intervention of the Ministry was much less, so there are no introduced soil conservation works and those that do exist are indigenous.

People and settlement

The people of both Andit Tid PA and Denqwan-Gunagunit PA belong to the Amhara ethnic group. They speak Amharic and their religion is Orthodox Christianity. There is no significant out-migration in either area. Since the rural land proclamation of 1975, land is owned by the government. Each household head has use rights in land which have been allocated to his family, and land can be transferred directly to the children during the lifetime of a family head. Land cannot be sold or hired by the farmers, a restriction which does not appear to affect the desire of local farmers to invest in the construction of ditches.

There is considerable socio-economic differentiation in the community in terms of the land owned and available labour. While the size of land holdings does not influence the practice of constructing ditches, differences in available labour do influence their timely construction. Household heads with large numbers of working-age family members can construct the ditches in a shorter space of time during final ploughing than a household head who has to do everything by himself or look to someone else for assistance.

Both areas have a long history of early settlement and agricultural cultivation. Owing to high population pressure (both human and livestock), intensive and continuous cultivation and relatively high rainfall, both areas have also suffered from an extreme degree of land degradation. According to the measurements by the Soil Conservation Research Project (SCRP), between 1983 and 1986 the average annual soil loss on cultivated land without soil conservation measures was 152.54t/ha, ranging from 77.94t/ha to 218.14t/ha (Yohannes, 1989).

New methods for old problems

Different types of soil and water conservation and afforestation activities are being carried out currently by the Natural Resources Development and Environmental Protection Department (NRDEPD) for Northern Shewa based on the farmers' full participation. However, no efforts have been made to study the different types of indigenous soil and water conservation techniques. Hence, no efforts have been made to improve the technical efficiency of the traditional ditches. In Andit Tid PA, from 1982 to 1992 the farmers were offered grain and edible oil by the Ministry of Agriculture as an incentive to build conservation works, such as different types of terraces, check-dams and cut-off drains on their cultivated land, and to construct feeder roads as well as to produce different types of tree seedlings. However, most of these activities, especially the physical soil conservation measures, have been destroyed by farmers. Most of the tree seedlings were planted on communal land so that the trees were frequently misused.

Generally, it seems that the introduction of new soil and water conservation as well as afforestation activities have failed to fulfil their planned objectives because the farmers did not participate fully at all levels of planning and implementation. Farmers were interested only in getting grain and oil, whatever soil and water conservation measures were constructed on their land.

Current trends

Farmers generally believed that crop yields had been decreasing because of the reduced size of land holdings owing to population pressure and poor soil fertility. Intensive cultivation, overgrazing, steep slopes and intensive rainfall have resulted in much of the topsoil being washed away. Thus, the soils are very shallow and less fertile over a large part of both areas. In order to maintain or improve soil fertility, farmers use traditional practices, such as crop rotation, fallowing, soil burning (known locally as *gay*) and to, some extent, manuring. According to farmers, if enough manure is applied every cropping season, the crop yield increases, although the use of manure as a fertilizer is limited as it competes with the use of dung for fuel. Farmers recognize that cereals give a better yield if they are planted after pulse crops. The average fallow period has also decreased to two to three years, owing to the shortage of cultivated land.

Gay is one of the methods widely practised to maintain or improve soil fertility. It is used mainly for the short rainy season (*belg*) crop, such as barley. It is believed that *gay* helps to increase yields, especially for the first year, when land treated with this method has higher yields than land which has been treated with chemical fertilizers. However, most of the farmers are aware that crop yields decrease rapidly if cultivated land is used continuously for more than a year. This indicates that farmers understand both the advantages and disadvantages of *gay*, which helps to increase the availability of plant nutrients such as calcium, potassium and phosphorus, while destroying organic matter in the soil (Tahal, 1988; Yohannes, 1989).

TRADITIONAL DITCHES: HOW THEY WORK

All farmers in both areas use traditional ditches on their cultivated land, irrespective of whether or not other forms of soil and water conservation measures already exist. Older farmers mentioned that they have been using traditional ditches for a long time and that many of them were inherited. For this reason, there are a lot of local farmers' sayings on traditional ditches: *Arso yalefesses temwagto yalewass* (from West Gojam), 'one can't imagine ploughing without traditional ditches as litigation without bail is unthinkable' and *Kezera gebere yaboyew yibelthal* (from North Shewa), 'a farmer who made traditional ditches is by far better than one who sowed'. These sayings indicate the historical importance of ditches within traditional farming practices.

Methods of construction and maintenance

Traditional ditches are constructed every cropping season using a *maresha* plough which is drawn by a pair of oxen after final land preparation, including sowing. On some difficult parts of the cultivated land, such as very steep slopes or at the bottom of terraces, hoes are used. Household heads (men) are mainly involved in the construction and maintenance of traditional ditches. A boy reaching his teens who is physically strong will also be involved after he has gained experience from his father. In some cases, other family members such as the daughter or mother are involved, especially to remove the loose soil. This decreases the amount of soil lost in surface run-off until the ditches are protected by crops.

Traditional ditches are not constructed on the same place on the cultivated field every cropping season. Farmers vary the position of the traditional ditches in order to avoid a gradual widening and deepening over time. Most of the farmers are aware of this risk. If the cultivated land with traditional ditches is put to fallow after the harvest, the ditch will not develop into a gully because it has already stabilized and the grass and other vegetation will cover it during the fallow period.

The width, depth, length, spacing and gradient of ditches differ from area to area. The width is determined by the width of the ox-plough (30–50cm). The depth is determined by the depth of the soil (5–25cm). The gradient of the traditional ditches is determined by each farmer and thus varies considerably (3–20 per cent). The spacing of ditches is dependent on the steepness of the slope, where steeper cultivated land has more traditional ditches than cultivated land on gentler slopes. However, traditional ditches are also used on almost flat and waterlogged land. The length of traditional ditches is determined by the length or width of each farmer's cultivated land. Although most of the farmers construct traditional ditches diagonal to the cultivated land, some are constructed perpendicular to the contour of the cultivated land. Therefore, in most cases, the length of traditional ditches is proportional to the length or width of the cultivated land.

Farmers construct different shapes of ditch based on their needs, ranging from linear ditches to contour ditches, arc ditches and cross-wise ditches. On waterlogged cultivated land the ditches cross one another for better drainage. However, the shape of the traditional ditches differs from one piece of land to another according to the interests of the owner of the land. On flat land where drainage is a serious problem (heavy clay soil), farmers use ditches with ridges, most of which are constructed in a diagonal or perpendicular position to the ridges. The ridges along the ditches do not prevent water movement because the excess water infiltrates easily and drains out of the cultivated land.

Different types of ditches are used on cultivated land which already has existing soil and water conservation methods. Some of the farmers construct ditches which drain the excess water to the side of the artificial or natural waterways. Others construct ditches which cut across the soil and water conservation measures – ie, bunds.

Farmers construct deep, wide ditches on the upper side of their cultivated land which is used as a cut-off drain to protect the field from the run-off coming from the higher land. On gentle and steep slopes, ditches are used to intercept and drain run-off to the artificial or natural waterways. On almost flat cultivated land, farmers construct traditional ditches in different directions to drain excess water (see photo 25).

Why are the traditional ditches built?

It appears that traditional ditches are generally used for two purposes:

1. To protect the soil from being washed away by run-off or overland flow coming from the uplands and to reduce surface run-off which is generated within cultivated land during the rainy seasons on steep and gentle slopes.
2. To drain excess water from almost flat cultivated land during the long rainy season.

Farmers also mentioned the advantages of traditional ditches, such as the protection of cultivated land from upslope run-off, the reduction of surface run-off and the minimizing of drainage problems on waterlogged cultivated land, which enabled them to achieve better yields.

Although no one mentioned the disadvantage of traditional ditches, there were cases of severe soil erosion owing to the improper use of the traditional ditches. Some of the farmers construct them perpendicular to the contour of the cultivated land with a higher gradient. Most of the farmers drain the excess water from their cultivated land to the natural or artificial waterways without considering the effect of this water on the stability of the waterways. Thus, the side-walls of the natural waterways or the artificial waterways are eroded by the water coming from the cultivated land. This will potentially result in the formation of big gullies unless some protective measures are taken by the owner of the cultivated land.

This lack of discrimination in the use of traditional ditches suggests that within a relatively short period (7–45 years) much of the Andit Tid area may be transformed into degraded land (Yohannes, 1989). Traditional drainage furrows (ditches) are not a useful physical conservation practice, but are a simple drainage technique which pose a severe erosion hazard on steep slopes with erodible soils (Tahal, 1988). In Dizi, Illubabor, level and graded drainage ditches are commonly used. It is not clear whether these ditches are considered conservation measures or are used to protect *teff* from waterlogging. On steep slopes – of up to 25 per cent – ditch spacing of 2m is common, which may also have a strong erosive impact (Hagmann, 1991).

Traditional ditches can thus contribute to soil erosion in both areas, particularly gully erosion along farm boundaries. The waterways are regarded as communal property and individual farmers do not pay any attention to the maintenance of waterways and gully rehabilitation. Sedimentation can also take place within the outside cultivated land, with traditional ditches carrying sediments away from cultivated land, although most of the farmers perceive sedimentation as a source of fertility, spreading accumulated sediments over the traditional terraces on their fields.

Nevertheless, many of the cultivated lands without traditional ditches are seriously affected by rill erosion. Even though no yield comparison has been made between cultivated land with and without traditional ditches, it does seem that the yields on land with traditional ditches are higher than they are on land which has no traditional ditches (Million, 1992).

CONCLUSION

The construction of ditches as an indigenous soil and water conservation technique is widely practised by farmers cultivating crops on the steep and often infertile slopes of the Ethiopian Highlands. Farmers perceive this practice to be an effective way of controlling soil erosion and waterlogging, and they build ditches regardless of whether other forms of soil and water conservation methods are introduced.

Despite their popularity, this study has also shown that the use of ditches can itself contribute to significant soil erosion, particularly if they are used regardless of the slope of the land or extent of existing degradation. Traditional ditches could be used properly by the farmers, particularly with regard to its gradient if they are found within a combined conservation system – for example, traditional ditches with graded traditional or modern, introduced terraces. This combination would force the farmer to construct the ditches with a lower gradient.

More studies are required to determine ways of increasing the efficiency of traditional ditches, and more information is required to compare the yield levels for the crops on cultivated land with and without traditional ditches. For farmers, the advantages of using traditional ditches nevertheless clearly outweigh the disadvantages, not least because the farmers themselves know and trust this method of soil and water conservation in the Ethiopian Highlands.

22

CREATING AN INVENTORY OF INDIGENOUS SOIL AND WATER CONSERVATION MEASURES IN ETHIOPIA

Hans-Joachim Krüger, Berhanu Fantaw, Yohannes G Michael, and Kefeni Kajela

The Ethiopian Highlands are home to many millions of small-scale farmers. In many areas, steep slopes and intense rainfall combined with the cultivation of unprotected soils have led to very high rates of soil erosion. Yet there is a wide range of indigenous soil and water conservation (ISWC) measures which farmers have developed over many generations, which can often provide the basis on which to construct improved systems of land husbandry. The inventory of ISWC measures described in this chapter provides detailed information about the diverse and ingenious ways which farmers have evolved to try to manage their land. The inventory demonstrates not only the strengths of these practices and the options for improvement, but also where major weaknesses exist.

BACKGROUND AND HISTORY OF SOIL CONSERVATION IN ETHIOPIA

The highlands of Ethiopia comprise 44 per cent of the country, include 95 per cent of the cropped area and two-thirds of the country's livestock. Approximately 88 per cent of the population live in this area, at an average density of 64 persons per km^2. In contrast, the lowlands comprise 56 per cent of the land area, but accommodate only 12 per cent of the population at a density of less than 10 persons per km^2. Most of the highland terrain has slopes of more than 16 per cent, and only a fifth is considered free from erosion hazard. Most of the productive topsoil in the Highlands has been degraded, resulting in chronic food shortages and persistent poverty. Serious erosion is estimated to have affected 25 per cent of the area, and some estimates find that 4 per cent of the Highlands are now so seriously eroded that they will not be economically productive again in the foreseeable future. The capacity of Highland farming communities to sustain production is therefore under serious pressure.

The Soil Conservation Research Project (SCRP) has estimated that about 1.5 billion tonnes of soil are eroded every year in Ethiopia. Similarly, the Ethiopian Highlands Reclamation Study (EHRS) estimated that between 1985 and 2010 the rate of land degradation will cost 15.3 billion Ethiopian birr, most of which (78 per cent) is due to crop failure or low yields. The Ethiopian government first recognized the severity of the soil degradation problem following the 1973–74 famine. With heavy external support, the government initiated a massive programme of soil conservation and rehabilitation in the most highly degraded areas. This involved the mobilization of peasant associations and the involvement of over 30 million peasant workdays per year (Hurni, 1986). Reports indicate that between 1975 and 1989 terraces were built on 980,000ha of crop land; 208,000ha of hillside terraces were constructed; 310,000ha of highly denuded land were revegetated (National Conservation Strategy, 1990). Yet these achievements fell far below expectations and, despite considerable efforts, the country is still losing an incredible amount of precious topsoil annually.

Environmental and socio-political factors have both contributed to this poor performance. Environmental factors include the dissected terrain, the cultivation of steeper slopes, erratic and erosive rainfall, and so on (Hurni, 1990; Campbell, 1991). Socio-political factors include the top-down approach adopted by bodies intervening to improve SWC. Farmers have been minimally involved in soil conservation activities, and indigenous knowledge has been undermined within planning, design and implementation processes. As a result, SWC programmes have to date proved to be highly unpopular among farmers. Government policies concerning land holding, marketing, pricing, credit and resettlement have discouraged long-term investment and exacerbated these deficiencies (Dessalegn, 1989; Harrison, 1987; Yohannes 1989).

INDIGENOUS SOIL AND WATER CONSERVATION IN ETHIOPIA

Over the last two decades many interesting and important facts have come to light about farmers' own SWC activities. Yet most experts are still largely unaware of the range of ISWC techniques used in Ethiopia. Part of the problem lies with the fact that there is no clear demarcation between indigenous and exotic (conventional or introduced) conservation measures. The definition and description of conservation measures has therefore proved problematic and is now a subject of intense debate.

Traditional conservation measures can be understood as farming practices that have evolved over the course of time, without any known outside institutional intervention and which have some soil conservation effect. Various mechanical, biological and agronomic techniques used by farmers in various combinations are incorporated within the term. These measures are the result of a gradual learning process and emerge from a knowledge base accumulated by rural people by observation, experimentation, and a process of handing down through the generations people's experience and wisdom. Traditional conservation practices are also shaped by and emerge from a detailed understanding of local conditions, and are modified in response to changing socio-economic, political and ecological conditions. Specific measures are generally used by farmers with a similar culture and within a discrete geographical area.

There are numerous definitional problems associated with conventional and indigenous conservation measures. For instance, many rural communities have been on the receiving end of introduced innovations (eg, the preparation of compost is an introduced innovation for some local communities in Ethiopia), but some of these conventional innovations have become deeply embedded in local practice. There are also problems concerning farmers' objectives for using conservation techniques. For instance, many constructions (eg, fences and walls) or agronomic measures (eg, manipulation of crop cover by thinning or the arrangement of cropping patterns) serve a range of purposes, of which conservation may be secondary, and farmers' awareness about and interpretation of these effects can differ enormously.

Indigenous conservation systems are often not only introduced to reduce soil loss or manage run-off, but are often designed to improve productivity and the suitability of land for cultivation. The order of priority associated with specific objectives may well change, and will involve modifying the design of the system or supplementing it with other measures. This reflects the evolutionary, dynamic and complex nature of ISWC systems.

There is a growing consensus that the poor record of SWC in Ethiopia can be attributed in part to the lack of appreciation of indigenous practices by soil conservation experts and policy-makers. Ethiopian farmers have long been aware of the problems associated with soil degradation, and have traditionally been conservation minded at farm level. However, the knowledge, skills, survival strategies and risks faced by farmers operating with low levels of external input have been ignored frequently by outsiders and experts promoting 'modern' conservation techniques. The result is that many introduced innovations have proved to be ill-adapted to existing systems. In recent years, a growing awareness of the limitations and hazards of conventional methods has encouraged experts to look to indigenous knowledge as a major untapped resource for developing sustainable agriculture.

Studies which identify ISWC measures, analyse their role in the context of local communities, systematize such knowledge, and identify the potential and limitations of such techniques, are therefore much needed. This chapter reports on a recent attempt to create an inventory of indigenous knowledge in representative parts of Ethiopia.

Creating an inventory of soil and water conservation

The objectives of this inventory are threefold:

1. To develop a sound methodology for the identification and description of ISWC measures, and the socio-economic, land-use and environmental context of the study areas.
2. To identify ISWC measures in selected study areas in Ethiopia: Armaniya in Shewa region; Konso and Gidole in Gamo Gofa; Anjeni in Gojam; and Hunde Lafto in Harerge.
3. To present findings, including important technical concepts and elements of the ISWC measures, and to assess their potential and limitations.

Many ISWC measures were identified during the few months' fieldwork. Detailed information on these measures can be found in the *Inventory of Indigenous Soil and Water Conservation Measures in Ethiopia* published by the Ministry of Natural Resource Development and Environmental Protection, Addis Ababa. The full results of this study, including the inventories, will appear in a forthcoming publication. In this chapter aspects of the inventory are illustrated using examples from Konso in Gamo Gofa region (see photo 26), where the main objective of the conservation system is water harvesting and soil fertility maintenance, and from Armaniya in Shewa region, where the system is aimed at the management of excess water during the rainy season.

The methodology

The development and testing of a methodological framework was a first stage in the formulation of an ISWC inventory system. The methodology is based on systems analysis and information is analysed at two levels. At the system level, this involves a description of the conservation system as a whole, and analysis of how different single conservation measures and effects within the system are related. At the level of single conservation measures, this entails a description of all single conservation measures which comprise the overall system.

The methodology takes a holistic approach by linking technical information

Sy	**Y/L/P/A/B** 2/0/1/0/0	**SUMMARY SHEET** **Indigenous Soil and Water Conservation Measures**
Name	Local Name: English Name:	*Kela* (Sidamo); *Yedengay Ereken* (Shewa); *Godeba* (Tigray) *Stone Bund*

Characterisation of the Measure

WATER

Type of Measure		Permanency		(Duration)
Agronomic Measure:	☐	Permanent:	☒	
Biological Measure:	☐	Semi-permanent:	☐	(5 -10 years)
Physical Measure:	☒	Seasonal:	☐	( )
		Shifting:	☐	( )

Purpose (Water or Soil Conservation)

SOIL

Water harvesting:	☒		Soil trapping:	☒
Water storage:	☐		Protection of soil surface:	☐
Water disposal:	☐		Soil improvement:	☐
Management runoff:	☐		Slope modification:	☒
Other reasons:	*Water conservation*			

Effect of the Measure: *reduction of slope degree & length; accumulation & protection of soil above & below the bund.*

Construction and Establishment

Materials

Kind of materials/plants used:	*Stones (medium to big size)*
Amount of materials/plants used per unit:	*1.0 - 3.5 m3/m*
Who designs the measure:	*Head of the family (male only)*

Labour

Source of labour:	*Family and neighbours*
Labour input per unit:	*3- 6h/m (secondary job between other activities)*
Organisation of labour:	*Within the family or neighbourhood*
Gender:	*Male only*
Time of construction:	*During slack season (February to March)*
From whom did the people learn the technique?	*From neighbours, ancestors*

Maintenance

Frequency of maintenance:	*Construction is permanent subject of development*
Indicators for maintenance:	*Overtopping of runoff during and after heavy rains*
Form of social organisation during maintenance:	*Family*
Source of labour:	*Family*
Labour input per unit:	*Not known*

DESCRIPTION OF THE ENVIRONMENTAL CONDITIONS

Morpho-logy

Typical physiographic position of the site:	*Upper to lower slope position*		
Range of slope degree:	*10 - 60%*		
Range of effective slope length:	*3 - 25 m*	Typical slope shape:	*Convex-concave*

Soil Characteristics

Texture	Top soil:	*siL/sicll*	**Soil depth**	Effective:	*> 45 cm*
	Sub-soil:	*sL*		Potential:	*> 150 cm*
Structure	Top soil:	*Sub angular blocky*			
Drainage	External:	*Good*	Internal: *Good*	General drainage class:	*Good*
Degradation	Main process:	*Nutrients losses, loss of organic matter*		Degree:	*Medium*
Soil moisture regime:		*Not known*		Surface stone content:	*0 - 20%*
Limiting soil properties:		*Soil depth, nutrient deficits (N,P), lower water storage capacity*			

Erosion Hazards of the treated spots (Micro level)

☒ **Water erosion**	Climatic factors:	*High & intensive rains*
	Morphology:	*Steep slopes*
	Soil factors	*Susceptible soils*
	Land use factors:	*intensive arable farming*
☐ **Wind erosion**	Critical period:	*Beginning of rainy season*

Erosion dynamic of the spot

Erosion disposition:	☒	Transmission position:	☒	Accumulation position:	☒

Description of indigenous soil and water conservation systems, Ethiopia

Erosion Status of the Treated Spots	Low: ☒ Medium: ☒ High: ☐
	Indicators of erosion status: *Low OM content in the top-soil*

Relation to the natural drainage system of the treated spots	Measure used at locations	within the natural water way: ☐
		outside but with influence on the natural water way: ☐
		outside without influence on the natural water way: ☐

| Description of the Measure

. Characteristics
. Function
. Application
. Land use type | *The leveled stone bunds are constructed along contours. The bottom width of the walls is 0.5 to 1m and the height about 0.4 (newly implemented bunds) to 3m (old bunds at the slope foot). Secondary stones and soil bunds cut through terraces, forming rectangular micro basins with the ridges.*
The Stone Bund measure is supplemented with agronomic measures (e.g. trash lines on the bunds, mixed intercropping, bushes and trees...). It is a permanent measure for water conservation purpose. The construction is carried out gradually, from existing bunds. The initial stage of the process is unclear due to existing alternative conservation techniques in Konso. |
|---|---|

Illustration (cut)	Illustration (front view)	Illustration (top view)

ASSESSMENT OF THE INDIGENOUS SOIL & WATER CONSERVATION MEASURE

Conservation Effects	Protection of soil surface: ☐	
	Reduction of slope length: ☒	*Stone bunds reduce the effective slope length.*
	Reduction of slope degree: ☒	*Stone bunds flatten the slope above the bund.*
	Reduction of runoff: ☒	*Increase of infiltration by reduced runoff speed; Enlargement of the water storage capacity of the soil.*
	Reduction of runoff speed: ☒	*This is the main effect of the Stone Bund measure.*

| Factors for the Acceptability of the Measure | . *Manageable with the available means of the farmers*
. *Supplements positive effects of other conservation measures*
. *Implementation by farmers assures best possible adaptation of the measure to local conditions*
. *Contributes to water conservation with significant effects on yields*
. *Positive effect on farming (improves water/nutrients balance)*
. *Relatively low labour input during slack seasons; presents no obstacle to the present farming system and traditions* |
|---|---|

Problems & Limitations	During establishment: - *slow implementation speed*
	During maintenance: - *maintenance is only of secondary importance*
	Current situation: *Systems must be supplemented by other agricultural measures to use the production potential in a more effective way.*

Options for improvement	. *Spacing of crops planted on the bunds could be improved.*

Feat for the Measure	. *Do political or administrative decisions/actions influence the use of the measure?* **Yes**
	. How? *Further investigation necessary.*
	. Tendency of applying the measure: *Implemented measures still under development; insecure land ownership jeopardises the long-term development strategy of the measure; food for work approach negatively affects the quality of the newly constructed bunds (due to quota-system); breakage of these poor quality new bunds causes severe damage to the conservation system.*

collected on site with ecological, socio-economic and farming system data. This approach helps to build up a picture of farmers' rationale for using specific measures. Local farmers provide much of the information, although general information about the research area is gathered from miscellaneous other sources.

Three different inventory forms are used to collect this information:

1. Form Y comprises all information about the system as a whole. The form serves three purposes: it lists all conservation measures undertaken within the conservation system; it looks at the linkages between single conservation measures; and it presents all additional general site information including the following:

   - Site location, including a map; location of a conservation system; environmental conditions, including the degradation status of the area; a description of the elements of a representative conservation system, including physical, agronomic and biological conservation measures. All the information is illustrated with a sketch map.
   - System management: system modifications over time, how local people organize for maintenance, whether the system has changed the feasibility of land use, farmers' perspectives on costs and benefits of the system, information on who benefits from the system; a description of the farming system.
   - Socio-economic information: demographic information, prevailing social hierarchies, existing social organizations, land-holdings; off-farm activities, farmers' perspectives on priority problems.

2. Form Sy comprises detailed technical information about single conservation measures.

3. Form Sys comprises short technical descriptions of conservation measures or effects.

General information relevant for the whole area is therefore presented once only in Form Y. This helps to avoid repetition.

Data collection therefore looks at various levels of the conservation system. For instance, a clear distinction can be made between the indigenous conservation system (eg, the water-harvesting system of the Konso area); single indigenous conservation measure (eg, the use of stone bunds within the Konso system); and a technical element of a single indigenous conservation measure (eg, filling with earth the spaces between the stones in a bund to allow fodder plants to grow). Some conservation systems consist of a single measure, unconnected with other measures. In this case, all information is entered into a single form (Form Ss).

The interpretation of information collected from local people by experts is a key issue. Information collected in the field is open to interpretation during analysis. The final results therefore need to be restituted to local people as a check against misinterpretation.

Some preliminary research findings

The research found that ISWC measures are practised in all areas of Ethiopia and are well adapted to local conditions. In general, plant cover is used to conserve soil in areas where vegetation has a high growing potential. In contrast, physical conservation measures play a key role in areas where vegetation cover is highly degraded or where there is limited growing potential for vegetation.

The research also identified the need to distinguish between those areas of the country with very well-developed indigenous conservation practices and areas where these are poorly evolved. Indigenous conservation knowledge appears to have accumulated particularly in those areas of Ethiopia where the natural resource base is under severe pressure from local communities, the ecosystems are fragile, and there is a long history of adaptation to adverse conditions. Traditional SWC techniques have been used widely for more than a century in regions like Tigray, North Shewa, Gamu Gofa, North Wollo and Harerge. In contrast, indigenous conservation measures are not well developed in areas of new settlement, where there is minimal pressure on the natural resource base. In these cases the introduction of conventional conservation measures may be more appropriate.

Some ISWC systems in Ethiopia have a wide geographical coverage (eg, traditional ditches in Armaniya), while others are confined to small areas only (eg, the stone bund and bench terrace system of Konso). Complex systems involving the use of many agronomic, biological and physical measures within one system tend to be less widespread than simpler systems.

ISWC measures are frequently site specific and accordingly vary in purpose. They may harvest water in lowland zones (with the help of trash-lines, tied ridges, level physical soil conservation structures such as *fanya juu* bunds); conserve soil *in situ* (traditional stone and soil bunds); dispose of excess water from crop lands during heavy rains (traditional ditches, cut-off drains); improve drainage (furrows and traditional ditches) and conserve soil while simultaneously increasing soil fertility (agroforestry, mixed cropping and intercropping). The research highlighted the fact that the success of many ISWC systems in Ethiopia depends on a combination of measures and effects rather than on a single technique (see photo 27). It showed that farmers should try to conserve soil and water with minimal inputs and achieve efficient conservation effects by using several measures together.

Field observations and interviews with farmers indicated the role played by patterns of land use in promoting soil conservation, such as rough ploughing, choice of crop and maintenance of crop cover. The choice of how different parts of the landscape are used is also important, and attempts were made to incorporate these issues into the description of local conservation systems. For instance, in Konso grazing is controlled on the upper slopes, plateaux and valley bottoms, which helps to maintain permanent vegetation cover in erosion-prone areas.

Although farmers in the study areas are concerned with both the short- and long-term benefits of SWC, more emphasis is placed on immediate or short-term benefits. Indeed, in the initial development stage of establishing certain measures, objectives other than conservation may be central to farmers' decision-making, while conservation aspects often become a dominant feature at a later stage. In Armaniya, for example, stone piles and heaps comprise the initial development stages of stone bunds. The stone content of soil tends to increase in areas with a high rate of erosion. Above a certain threshold, the accumulation of stones can seriously impede cultivation and reduce crop production. Farmers collect stones and create a pile of a maximum of four stones. The second stage in the creation of stone bunds is the transfer of stones from these piles to semi-permanent heaps. This work, undertaken generally during the slack season, can have positive yield effects and reduce subsequent labour inputs to cultivation. After some years, farmers convert these heaps into stone bunds. By reducing slope degree and length and allowing the accumulation of soil above the bund and the protection of soil below, bunds comprise an important part of the physical conservation system in Armaniya. This example also shows that conservation systems are derived from small seasonal inputs over a long period rather than large inputs at any one time.

The effect of restricting an inventory to descriptions of conservation measures would be to overlook the rationale behind people's use of these measures. Recognizing initial measures, even if they themselves do not directly contribute to conservation, is vital to understanding how the conservation system as a whole has evolved.

Lack of uniformity among similar ISWC measures is common. For example, stone terraces are widely used by farmers, but the design and function of terraces vary considerably within and between households. Some stone terraces are continuously built up and developed into bench terraces, while others are dismantled every two to five years to redistribute the soil accumulated behind the bund. Detailed investigation of the same indigenous conservation measures in different locations during the research identified specific technical elements used in a few places only. In Armaniya, an area where land is scarce and where yields are decreasing and unstable, physical conservation structures occupy a significant share of productive land. This can result in serious conflicts between longer term conservation and immediate food needs. However, farmers successfully convert unproductive stone bunds into productive areas by cultivating the space between the stones, which are then filled with soil and planted with perennial plants like *gesho*. Dry *gesho* leaves are used for beer brewing. In this way conservation and income generation can be combined.

The above example is indicative of the fact that conservation measures tend to be more acceptable to farmers if they serve multiple objectives and help to increase production. Indeed, to many small-scale farmers resource conservation cannot be an end in itself, but is an integral part of efforts to improve and sustain livelihoods. The research found that improving productivity is the underlying rationale which cuts across all the agro-climatic zones inventoried. Whether the function of a specific ISWC measure is water harvesting using trash-lines and tied ridges in dry zones (as in Konso, Armenia and Gidole) or run-off control by stone bunds (Armenia), or traditional ditches and cut-off drains (Gojam, Debre Birhan) in high rainfall areas, the overriding concern of Ethiopian farmers is yield levels. As with the cultivation of *gesho* on stone walls in Armaniya, stone bunds in Konso are used productively to grow field crops, including legumes, safflower, sunflower and fodder plants. Appropriate ISWC measures therefore contribute to the diversification of activities and the maximization of choices within livelihood strategies, thereby minimizing risk.

The maintenance and improvement of soil fertility are important aspects of the conservation concept. In Konso people collect all organic material from homesteads, including manure from the stalls and vegetable waste and ash from cooking, and store it for several months in specially prepared pits on village boundaries. The compost is then transported to the fields at the beginning of the growing season and applied to degraded spots or to highly responsive crops such as maize or sorghum. Farmers are most interested in the yield increase, while conservation effects, such as improving infiltration within the soil, are of secondary importance.

The development of indigenous conservation systems is governed by individuals or small groups, and so ISWC activities are mainly associated with the lowest planning levels. In contrast, the planning and implementation of such systems on sub-catchment or catchment level is in principle the responsibility of large social structures (eg, water distribution committees). Recognizing the potential role played by these institutions in ISWC is vital. It is also very important to understand the linkages between the various planning levels (plot, farm, sub-catchment, etc). In some cases, farmers invest in conservation measures outside their immediate farm area in order to avoid damage to their own farmland. However, many farmers cannot afford to invest much time and energy in conservation activities aimed at improving natural resource management in areas outside their immediate and direct interest. In some cases, therefore, gaps exist between different conservation systems operating at different levels. The absence of an effective local organization to deal with resource management and planning at levels higher than the farm reinforces this problem.

CONCLUSION

Overall, the research found that there is a large pool of ISWC measures in Ethiopia to be tapped, strengthened or modified. This inventory of measures has produced information on a wide range of interesting aspects of these systems, which can guide researchers in their identification and development of efficient and acceptable technologies. The research also suggests that there is plenty of scope for incorporating certain aspects of ISWC into conventional approaches, and vice versa, to improve their acceptance and appropriateness. ISWC therefore has considerable potential for use in the development of new approaches for working with farmers.

A broader interpretation of the term 'conservation measure' needs to be adopted to include those modern farming practices which enhance conservation farming. However, modern measures should form part of a broader strategy based on an in-depth understanding of the prevailing farming systems and their dynamics. The promotion of individual conservation practices in isolation is unlikely to have a lasting impact, especially if the focus is on their technical advantages rather than on their overall compatibility with local farming systems.

23

LOCAL FARMING IN THE FORMER TRANSKEI, SOUTH AFRICA

Kevin Phillips-Howard and Chris Oche

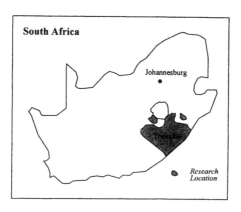

The Transkei is one of South Africa's former homelands, located in the north-eastern part of what is now Eastern Cape Province. It remains a poor, densely populated and comparatively undeveloped region of the new South Africa. Government approaches towards improving agriculture to date have been 'top down'. There has been little investment in small-scale subsistence farming, yet as this paper shows, farmers have experimented with a range of local soil and water conservation (SWC) techniques. As pressure on resources increases, these indigenous methods will become more important in developing a sustainable resource base for the Transkei (see photos 28 and 29).

INTRODUCTION TO THE AREA AND THE PEOPLE

The Transkei lies between latitudes 30°S and 33°S and longitudes 26°45'E and 30°15'E in the wetter, eastern region of South Africa, where most of the normal annual rainfall occurs in the summer between October and March. The rainfall in this part of South Africa is known to be comparatively reliable, with a long-term average of 815mm per annum, varying from less than 750mm in the south-west to 1400mm along the north-east coast of Transkei. The climate is generally humid sub-tropical, but increasingly *montane* inland.

Generally, the Transkei comprises a series of step-like land surfaces. It is dominated by the coastal and inland plateaux which are separated from each other by a pronounced and deeply indented escarpment. The coastal plateau runs parallel to the coast, rising gradually from 600m above sea-level to 700–800m at the boundary with the inland plateau. The inland plateau, at about 1250m above sea-level, comprises a belt of coarse, pebbly sandstones separated by fine-grained sandstone, mudstone and shale. This plateau varies greatly in width as its seaward escarpment is extremely indented by vast gorges with prolonged spurs or tongues. Above and behind the inland plateau towers the basaltic escarpment of the Drakensberg Mountains, which rise to 3000m above sea-level. Much of the area thus has slopes greater than 5 per cent and at least 60 per cent of the land has slopes exceeding 15 per cent (McKenzie, 1984).

The people of Transkei predominantly speak Xhosa, an Nguni Bantu language. Their traditional subsistence base has been agropastoralism supplemented by hunting, gathering and, near the coast, fishing, including the collection of shellfish.

There are now in excess of 3.5 million people living in Transkei. This figure would be considerably higher if it was not for the high level of out-migration from the region. Around 1 million (predominantly) men between the ages of 20 and 60 years leave the Transkei to work elsewhere in South Africa. Nevertheless, population density remains high at around 84 persons/km^2, with the lowest densities in the rural interior and the drier or more mountainous western areas.

A brief history of land use in the Transkei

Transkei was scheduled as a Native Reserve for exclusive African occupation as early as 1913 and Xhosa-speaking and other Africans were forced to move there from areas reserved for white occupation. Within the context of the overall policy of apartheid or 'separate development', the government opted for the establishment of 'betterment schemes', which were essentially exercises in physical planning, emphasizing land rehabilitation. These schemes involved replanning villages in regular grid-iron patterns and allocating standard houses and garden plots in defined areas or wards. Over a 30-year period from about 1945, some 70 per cent of the total area of South Africa's former homelands was subject to such 'betterment'. The main effects of these schemes from the tribal leaders' viewpoint were largely negative, resulting in increased economic hardship and agricultural underdevelopment, a loss of local autonomy, a breakdown of established tenure systems and deteriorating ecological conditions with an accompanying loss of local environmental knowledge (McAllister, 1992).

The last initiative of the 'independent' government was the 1991 Transkei Agricultural Development Study which aimed to replace subsistence farming with commercial production, especially of beef and cattle. This scheme in particular ignored how rural Transkeians actually use their land and livestock, and relate to each other and the natural environment.

The most recent soil conservation initiative to take place is the Transkei Soil Conservation Programme (TSCP). This programme involves the construction of conservation structures such as stormwater controls to channel water in a non-destructive way towards natural drainage channels. In these channels, weir structures built of stones, gabions and brush are used to control erosion. Reeds and grasses (in particular, vetiver grass) are then planted in the silt to stabilize the water. The TSCP is deliberately designed to employ as many people as possible in order potentially to ensure its long-term success. Although the programme is reported to be successful, it remains to be seen if this can be sustained. Will the local people continue to maintain the soil conservation structures indefinitely without payment? Or will they, as happened during previous initiatives, abandon or adapt these structures once the incentives are removed?

Overall, there has been little government investment in the small-scale farming sector in Transkei. Consequently, the area continues to be characterized by sub-subsistence level agriculture which has declined markedly over the past century or so. Cattle, which are communally grazed, and maize are the principal subsistence products. Crop yields are low and livestock productivity is far below its potential. The average family possesses six head of cattle and a large number of smaller stock. At least 40 per cent of rural families in Transkei live in a state of poverty and possibly one-third have below minimum energy intake (Bembridge, 1987a). Farming has been left increasingly to women, old men and children. For substantial changes to occur, a vast investment is needed to overcome the economic dependency and underdevelopment of the Transkei (Southall, 1982; Beinart, 1982).

The evolution of farming in Transkei

Traditionally, a system of shifting cultivation was practised over much of the Transkei, the land regaining its fertility as a result of being left to lie fallow. This system was well adapted to the exigencies of climate and soil, and provided subsistence with minimal labour input. The crops grown included sorghum, which was progressively replaced by maize throughout the 18th and 19th centuries, pumpkins, gourds, melons, wild peas, beans, cocoyams, guavas, mangoes and citrus. Milk rather than grain was the staple food (McAllister, 1992). Tobacco was also increasingly grown as a trade item by the end of the 18th century. A range of edible tubers, roots, berries and leaves were also collected in times of shortage. This was supported by much local environmental knowledge (Bundy, 1988). During the colonial period, this traditional system was undermined and, as elsewhere, Transkeian farmers were forced to maximize their use of land and to skimp on conservation measures.

By the early 20th century, plough agriculture was expanding over greater areas, which were cultivated more thoroughly and perhaps more carelessly than in hoe agricultural systems. Together with land alienation and population growth, these changes imposed great pressure on the available resources. Despite increases in grain and livestock production up to the 1930s, this pressure, together with overgrazing, land deterioration, increasing dependence on migrant labour, competition from white farmers, lack of inputs and support, and natural calamities such as droughts, resulted in decreased agricultural production (McAllister, 1992). The system that evolved depended on migrant labour to provide access to oxen and ploughs, and the collaboration between homesteads to form work parties at certain times.

From the 1940s population pressure resulted in land shortage, grazing areas were reduced, fallows were cut and more marginal land was brought under cultivation. The resultant decline in soil fertility, together with reduced homestead size and shortage of draught animals, prompted farmers to abandon their fields and to concentrate more on gardens adjacent to their homesteads (McAllister, 1992).

This represented a significant change in the farming system. Gardens were easier to work, as they were more accessible and less labour-demanding. They were also easier to manure and to protect, as they were fenced and close to the byre. In addition, gardens were more amenable to intercropping, which helped to control erosion and to maintain soil fertility; they could also be enlarged or moved to include a fertile old kraal site. Thus, gardens became an important element of the farming system. For many farmers they became more important than the fields (McAllister, 1992).

The farming system in Transkei is closely linked with the system of land tenure. Most of this area is subject to a modified form of traditional land tenure in which the land belongs to the state and is allocated through the tribal authority. Rights to communal grazing remain for those who have livestock but, owing to population growth and land shortage, the customary right of all married men to be allocated a field can no longer be met (Hawkins Associates, 1980).

Currently, maize is grown on about 500,000ha. This accounts for about 88 per cent of the land under food crops, while sorghum accounts for 7 per cent and legumes and vegetables about 5 per cent (Bembridge, 1987b).

The livestock population of the Transkei has apparently remained static over the past 30 years with around 1.5 million cattle, equivalent to about 0.5 cattle per person. However, local and regional differences in agro-ecological conditions are reflected in variations in the Transkei's farming system. These range from intensive specialized production and mixed farming, to semi-extensive mixed farming (Hawkins Associates, 1980). For example, in the coastal areas there has been more emphasis on maize, other subtropical crops, fruit and vegetables. In the more humid area, agroforestry and mixed cropping were highly developed and production met subsistence needs, which was unusual for the Transkei (Beinart, 1982). Also, in the more humid area, higher-than-average cattle densities occur. Recently, sugar and tea monocultures have been introduced to the area. Further inland, the system becomes less intensive with declining rainfall, and the role of cropping becomes more limited.

THE IMPORTANCE OF SWC METHODS IN THE TRANSKEI

Around 25 SWC techniques were identified in a survey which took place in over 12 locations spread among the three main agro-ecological zones of Transkei, which include the mountainous interior, the central undulating plain and the rugged coastal strip. This is a surprisingly large number, given the marginal and unproductive nature of agriculture in the area. The table below shows the distribution of the techniques in terms of reported presence in a location and the number and proportion of farmers using them. Some techniques, like water harvesting, drainage ditches, tied ridges, valley farming and planting pits, appear to be spatially limited and, where present, used by a minority of farmers. The focus here is on the five most widespread and frequently used techniques: contour ploughing, irrigation, mixed cropping, crop rotation and manuring.

Earth contour bunds

Within the main fields, and sometimes the gardens too, ploughing and planting are often carried out along the contours. Typically, fields are laid out in strips which vary in width, being narrower on steeper slopes, and are often 250–500m

Distribution of soil and water conservation techniques

SWC technique	Presence in locations	Percentage of farmers
Terraces	6	12.5
Control of gullies	6	10.8
Earth contour bunds	12	60.8
Water harvesting	3	5.8
Drainage ditches	3	5.0
Crop mixtures	10	36.7
Tied ridges	3	5.0
Valley farming	2	2.5
Irrigation	10	37.5
Planting pits	3	2.5
Vegetation barriers	7	15.8
Stone lines	4	9.2
Dams	5	18.3
Crop rotation	11	42.5
Tree protection	7	20.8
Mulching	8	15.0
Manuring	11	64.1
Others	4	13.3

long. In between these are uncultivated contour ridges or banks accompanied by ditches. The bunds are usually around 0.5m high and 2m wide and covered with grass. They are usually made of earth which has been turned out of the ditches' downslope. Grasses are sometimes planted or are encouraged to grow by placing cut grass as mulch along the bunds. Along the upper side of the cultivated area, cut-off drains are excavated, which are large earthen banks designed to intercept overland flow.

Contour bunds are usually constructed by tractor and plough. In the past this work was often carried out by the government, particularly in connection with 'betterment' schemes. The amount of labour involved varies from 1 person-day working with a machine to more than 30 person-days, where a team of 5 or 6 people do the job by hand over a period of up to 6 days. Maintenance of the ditches and bunds is the responsibility of the farmers and their families. In many areas, maintenance is poor or non-existent, possibly because of the high labour requirement.

Since contour bunds were introduced by the government as a conservation measure in the early 1930s, farmers have been experimenting with them in various ways:

- Destroying them because of perceived wastage of space.
- Reconstructing them downslope to release accumulated soil fertility in the bunds.
- Planting bunds with grass strips or placement of mulch to encourage the growth of grass.
- Growing fruit trees along bunds where protection from livestock is possible.
- Reducing the width of bunds to save space.
- Modifying the bunds, especially where rocks and other impediments prevent their normal construction.

While the design of these bunds has remained the same, the main point is that farmers regard them very much as temporary structures. Destroyed bunds are rebuilt downslope, generally every two to three years.

The prospects for contour bunds thus seem to be ambiguous, for although this is a widely established practice with recognized conservation and economic advantages, there is also evidence of inadequate maintenance of these conservation structures (see photo 30). However, the adaptation or indigenization of the technique may enable farmers to increase the advantages and decrease the disadvantages associated with contour ridging.

Irrigation

The main sources of water for irrigation are surface sources: streams and rivers but also some springs. Access to water is not subject to much regulation. Most of the farmers practise manual irrigation using watering-cans, tins, hoses, plastic bags and gravity-fed systems using underground pipes and cut channels. In most vegetable gardens, crops are planted on flat beds and irrigated by hand, by both men and women. By enabling crops to be grown in the dry winter season, irrigation can supplement the rainfed crop harvest.

Irrigation in Transkei appears to be primarily subsistence-oriented, although a few farmers produce irrigated crops for the market. The areas irrigated by farmers averaged 0.38ha, equivalent to the size of a typical garden plot. Larger scale market-oriented irrigation was identified on plots ranging in size from 2.5 to 19ha, all of which were individually managed and located in valley bottoms. The main crops grown by irrigation are cabbage, spinach, beetroot, carrot and tomato. However, in the moister coastal zone, fruit trees such as bananas, avocados and oranges also benefit from irrigation. Maize crops are not irrigated.

The majority of farmers said that they began to irrigate at least 25 years ago or that the practice originated with their forefathers. Although irrigated agriculture has long been promoted in Transkei for commercial and welfare purposes, it has seldom proved to be economically viable. The few successes have included older men with multiple businesses, farms adjacent to large schemes, and a well-organized purchasing and marketing co-operative (McIntosh et al, 1993).

The variety of irrigation techniques practised in Transkei enables all farmers with access to water to irrigate, at least using a bucket, watering-can or tin. The factors constraining commercial irrigation were identified as the lack of skilled labour and the need for constant checking and supervision of work. This reflects a difference in perspective between subsistence-oriented and primarily commercial farms which require a significant investment in equipment. The land constraint is manifest in an absence of water regulations, lack of access to irrigable land and inadequate assistance and farmer support from within Transkei (McIntosh et al, 1993).

Generally, the potential for irrigation in Transkei is comparatively good because of the relatively high rainfall and the presence of perennial streams and rivers. Small-scale irrigation is associated with subsistence needs and is likely to endure, despite the absence of institutional or policy support. Irrigation in gardens could be further promoted through community-based projects that involve more development of water sources, appropriate training and extension, and research and development on labour-saving techniques. The expansion of commercial irrigated farming, on the other hand, requires many constraints to be overcome, including access to land, capital, inputs and appropriate knowledge and skills.

Mixed cropping

Mixed cropping occurs in both the homestead vegetable gardens and the main maize fields. It has multiple functions, including soil conservation, spreading labour, harvest and risk. In the main fields, beans and pumpkins are often intercropped with maize. These may be planted by hand or using a mechanical planter drawn by oxen or by a tractor. Pumpkins are frequently planted by hand in between the maize plants.

From pre-colonial times onwards, various vegetables such as pumpkins, gourds, beans and peas, as well as tobacco, were grown in gardens close to the homesteads. Traditionally, these were cultivated by women and subdivided by crop. In the coastal zone, more root crops and trees, such as guava, mango and some citrus, were included in an agroforestry system. The displacement of sorghum and the wider spacing of maize has made the system more amenable to mixed cropping. The mixture of beans and pumpkins has become an integral part of the system. The practice has evolved mainly through the incorporation of more crops into the farming system – notably, maize mixed with beans and pumpkins in the main fields, and new vegetables (cabbage, spinach, onions, carrots, etc) in the homestead gardens. The changes in practice towards mixed cropping in the cereal fields may be associated with the labour savings that can be gained for equal crop output. This is particularly evident with pumpkins, which reduce the need to weed (Beinart, 1982). The current pressure to grow beans in sugar-cane plots reflects a concern among women to produce food for the family as well as cash crops on scarce land (Porter and Phillips-Howard, in press).

Although mixed cropping could be threatened in the future by government promotion of monocropping, it is likely to continue. None the less, this practice could be reinforced by the development of labour-saving techniques, especially to help with weeding.

Crop rotation

Crop rotation occurs in both the homestead gardens and in the main fields. It has long been part of Transkei's farming system, evolving from a simple form mainly involving cereal crops in the main fields, to a complex rotation involving leaf, root and pulse crops.

Apart from the sequences of cropping and fallowing in the main fields and gardens, crop rotation currently appears to be most important in the homestead gardens where cereals, and especially vegetables, are interchanged. One specifically mentioned sequence, which takes place with irrigation and in the more humid coastal zone, involves interchanging leaf crops (eg, cabbage and spinach), root crops (eg, potatoes, beetroot, carrots and Indian potatoes) and pulse crops (beans and peas). Crop rotation in the form of occasionally shifting the position of the homestead garden also occurs. Where sorghum is still grown, this crop is interchanged with maize which grows well in soils that were previously planted with sorghum (Beinart, 1982).

It seems that crop rotation will continue to occur in the homestead gardens and because of its acknowledged advantages (especially higher yields, and soil and water conservation), it may endure in the main fields, too. However, research and development work may be needed to promote this technique through attention to the disadvantages of weeds, labour and applicability under drought conditions.

Manuring

Manuring, including the application of chemical fertilizer, occurs in both the homestead gardens and the main fields. Planters are frequently used which simultaneously drop seeds and pour manure. Kraal manure was mentioned by most of the farmers who use this technique. Over half also used inorganic fertilizer which is usually mixed with kraal manure. A small percentage also reported that they use compost.

Manuring has become more important with the reduction of fallowing and increased pressure on the land. Many farmers concentrate more on the gardens than on the main fields. As agropastoralists, Transkei farmers have long combined livestock husbandry with crop production, including the use of manure as fertilizer. Traditionally, crops were often planted on the downslope side of the kraals in order to receive manure through run-off. Manure would also have been taken to other nearby plots and those established on the sites of old kraals. The more distant sorghum and, later, maize fields probably relied more on fallowing and crop rotation for the replenishment of fertility. Although livestock numbers may have remained constant, manuring has increased, especially the use of inorganic fertilizer in combination with kraal manure. Some farmers suggested that the mixture of kraal manure and inorganic fertilizer produced better results than either used alone.

Manuring will undoubtedly continue because, with increased land pressure, there is no other means by which many farmers can fertilize their crops. However, the amounts applied may be inadequate, despite increased usage (Bembridge, 1987b). In this case, promotion of the technique will require problems of access and supply to be overcome. Research and development needs to be directed at labour-saving alternatives as well as at farmer-led experimentation with mixtures.

CONCLUSION

Overall, the approach to SWC in Transkei, as manifest in 'betterment' schemes, the policy recommendations in the 1991 Transkei Agricultural Development Study and the TSCP, has been authoritarian or 'top-down'.

Despite the relative failure of such schemes and the lack of any institutional or policy support for small-scale agriculture, farmers have developed a number of widely established conservation practices which have enabled them to gain higher crop yields from limited areas while conserving soil and water. It is likely that farmers will continue to adapt and develop such indigenous SWC methods.

The introduction of the new South African Government's Reconstruction and Development Programme (RDP), with its emphasis on sustainable agriculture for small-scale farmers, offers better prospects than ever to incorporate such indigenous techniques into agricultural development in Transkei. The RDP aims to improve the quality of life by 'efficient, labour-intensive and sustainable methods of farming' (African National Congress, 1994; p 104). Such methods of farming are likely to include indigenous SWC techniques, such as those described in this paper, which have been retained by farmers in Transkei against all the odds.

24

TRADITIONAL SOIL AND WATER CONSERVATION TECHNIQUES IN THE MANDARA MOUNTAINS, NORTHERN CAMEROON

François Hiol Hiol, Dieudonné Ndoum Mbeyo and François Tchala Abina

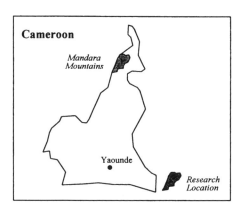

An intensive system of soil and water management has been developed by local people over the centuries in the Mandara Mountains of Northern Cameroon to restore and maintain soil fertility. Terraces, rows of stones, small dykes, irrigation and drainage canals, wells and micro-dams are combined with a range of biological methods, such as agroforestry, composting, mulching, growing complementary crops and crop rotation. Over the last three decades, improved security and service provision have encouraged people who were originally settled on the slopes of the Mandara Mountains to migrate to the adjacent plains, leading to changes in soil conservation techniques in both the mountain and the plain.

INTRODUCTION TO THE AREA AND PEOPLE

The Mandara Mountains run along the Cameroon-Nigeria border and reach heights of 1200 and 1500m, then fall through a series of plateaux to the surrounding plains below. The climate is characterized by alternating dry and wet seasons. The rains last from April to October, the wettest period being July–August. Rainfall is encouraged by the presence of the mountains. Annual variations are very marked and there are also significant differences from place to place. Mokolo, the weather station at the highest altitude in the survey (800m), had an annual rainfall range over the period 1985–93, from 784 to 1193mm, while for Mora, at the lowest altitude (440m), annual rainfall over the same period varied from 408 to 751mm.

Soil and vegetation

In the mountains and plateaux regosols are found in juxtaposition with rocky outcrops and lithosols. The most common rocks belong to the 'base complex' and are either acidic or mesocratic. In the foothills, at 400–600m altitude, soils are made up of colluvial deposits. The source matter is a thick accumulation of quartz, feldspar and mica grains which are soft, coarse and permeable. These soils are currently threatened by gully erosion which is removing the layers by drying out the soils and creating an uneven relief. This terrain is considered marginal for rainfed crops.

Soils of the valleys and plains are made up of clayey sand mixed with gravel. Most of the light- or medium-textured soils regenerate well, even though fallow periods are quite short. Although structural degradation, combined with increasing impermeability and gully erosion, has been noted overall, these are high-quality soils, producing a variety of crops – cotton, sorghum, millet and groundnut.

Population

The Mandara Mountains are one of the most populous regions of Cameroon. According to the 1987 census, there were 753,583 inhabitants, giving an average density of 84 inhabitants per km^2. But the population is unevenly spread; in the north, the population exceeds 100 per km^2, whereas in the south, the average is 40 per km^2. Population growth is estimated at 2.6 per cent per year.

The major ethnic group of the Mandara mountains is the Mafa, but there is great diversity, with the Kapsiki, Bar, Daba and Fali in the south, and the Mandara, Peuls, Podoko, Hidé, Ouldémé and Mafa in the north. The Mafa are concentrated to the north of Mokolo, where in some areas densities rise to over 300 inhabitants per km^2.

Since the colonial era, the authorities have made efforts to induce the mountain dwellers to come down to the plains, but they have always resisted this pressure. Recently, however, the combined effects of demographic pressure in the mountains and the prevailing stability in the region, as well as social initiatives carried out by the government and NGOs in the plains (schools, health centres and water supplies) have led to migration from mountain to plain.

There are two types of migration: temporary and permanent. In the case of permanent migration, whole families have abandoned their lands to make a new life for themselves elsewhere. In the case of temporary migration, a few family members leave for a period rarely exceeding two months, in order to look for paid work.

Land ownership and use

Land is held under private ownership, having been acquired by either clearance or transfer, but clearance of land not previously occupied is rare nowadays because of the scarcity of unoccupied land. Only male descendants can inherit land. In cases where the deceased has no male heir, the land reverts to his brothers, cousins or nephews in the paternal line. Among the sons there is always a privileged heir, who may be the eldest or the youngest son, depending on the ethnic group. He is the one who inherits the paternal home and, among certain groups, also has greater entitlement in the division of the land.

Sale of land among the Mafa is rare and occurs generally between members of the same clan or family. Gifts are more common than sales among the Mafa and generally happen between friends or relatives. The temporary transfer of land can take two forms: loan or lease. Leases are more frequent among the Mafa. The rent is fixed according to the fertility of the land and its manageability and is paid either in kind (goats, sheep, millet) or in cash. Annual rents vary between 12,000 and 24,000 Fr CFA per ha (equivalent to US$16–32).

On the plains at the foot of the mountains, cotton is the main cash crop, while sorghum, groundnut and cow peas provide food and fodder. A cotton/sorghum rotation is practised, into which groundnut may be introduced if the soil is not too heavy. In some places land is under permanent cultivation although, where possible, land is fallowed for three to five years. Land preparation is carried out by plough as well as by hand. Organic manure is little used, although during the rotation period plants benefit from the residual fertilizer provided by the Société de Développement du Coton (SODECOTON) for cotton growing.

The average farm has a couple of cattle, four to ten small stock and one or two donkeys. These animals are kept for ploughing, transport, manure production and for the income supplement they generate when sold; to a lesser extent, they provide milk and meat. During the growing season, animals are looked after by children and graze in the fallow fields and uncultivated areas, then are penned around the plots in the evening. In the dry season, they wander around freely, grazing on the crop residue as well as in the uncultivated areas. Farmers with larger herds entrust their animals to professional herdsmen.

In the mountains and plateaux the main crops are sorghum, pearl millet, cow pea and groundnut, with small quantities of sesame, okra and dah (*Hibiscus* spp). Various pumpkins, tobacco and yams are grown near the huts. Taro is almost always grown in moist areas and near running water. Groundnut is the principal cash crop with small amounts of cotton.

Organic manure, the only type of fertilizer brought to the fields, is obtained from various sources: animal and domestic waste, different types of compost, litter from the trees remaining in the fields, and ash resulting from stubble-burning. The preparation of fields for cultivation entails cutting overhanging branches and pollarding large trees. Useful waste is gathered up and the rest is burnt. At the same time, terraces are maintained and terrace walls repaired. The land is not ploughed before sowing, but during the growing season the ground is hoed and earthed up twice – the first time two weeks after sowing and the second time between June and July.

In the mountains, livestock rearing is intensive, playing an important religious and social role (eg, sacrifices and feasts) and provides a form of investment to be sold in case of an emergency.

Constraints to production and local strategies

For the last ten years there has been a reduction in rainfall, which, combined with its uneven distribution, has brought low yields. Mountain soils have been kept fertile thanks to the regular application of organic manure, whereas the soils of the hills and surrounding plains have decreased rapidly in fertility.

Water erosion is very marked in the plains and hills, mainly in the form of rill and gully erosion. This is chiefly linked to the fact that most of the soils in this area are unevolved, colluvial soils. Erosion has been aggravated by the clearance of marginal land, the spread of animal traction and the cotton cultivation system.

In the mountain areas, agricultural production is only possible through the regular maintenance of terraces and the intensive management of soil fertility. Natural resources in this area are in acute demand owing to population growth. The most serious problem encountered by farmers in cereal production is the presence of *striga* (witchweed). Infestations are most severe in the plains and the plateaux, in fields exhausted by continuous cultivation and in the short fallow periods. Caterpillars and fungi are also a problem in certain villages.

Agricultural credit is needed to buy fertilizer, selected seeds, pesticides, draught animals and equipment. But access to credit is very limited. The Mandara Mountains appear to have a huge potential for groundnut production and for other crops (taro, yam, potato, manioc), which are presently of only secondary importance. However, marketing structures for these crops are currently non-existent.

So far as livestock are concerned, in the mountains, the search for fodder is very labour intensive and, during the dry season, long journeys are necessary to find fodder and water (both for people and livestock). In the surrounding plains and plateaux, the main constraint is the shrinking of available pastures.

SOIL AND WATER CONSERVATION (SWC) TECHNIQUES

Physical structures

The terrace (*medoedoe*) is an edifice which comprises two parts: the low wall and the bank (see the figure below). The supporting wall, the foundation of the terrace, is made up of different-sized dry stones which are arranged in a compact manner in order to maximize the overall density of the wall. It can either be vertical or inclined towards the crest of the mountain. Its height varies between 25 and 200cm, and its thickness decreases progressively from the base to the summit.

The bank is the part of the terrace which holds the crops and varies in width from 40 to 190cm. It is either horizontal or slightly inclined against the slope of the land. The object of this is to retain as much water as possible on the bank. The Mafa distinguish three parts to the terrace, namely:

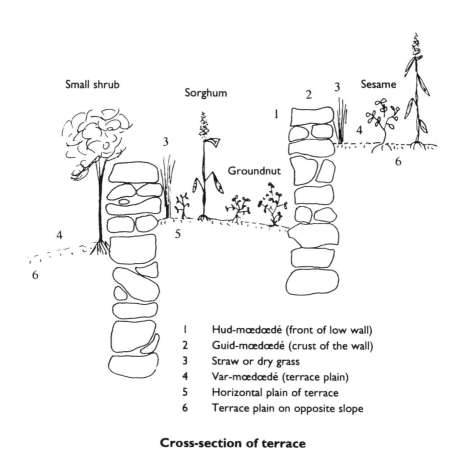

1	Hud-mœdœdé (front of low wall)
2	Guid-mœdœdé (crust of the wall)
3	Straw or dry grass
4	Var-mœdœdé (terrace plain)
5	Horizontal plain of terrace
6	Terrace plain on opposite slope

Cross-section of terrace

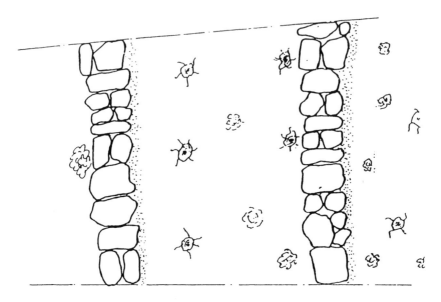

Aerial view of terrace

1. The *hud-medoedoe*, the main façade of the wall (seen from the front).
2. The *guid-medoedoe*, or the crest of the wall.
3. The *var-medoedoe*, or the cultivable part of the terrace. They also identify two types of terrace, the mountain *medoedoe* and the *inagleam* of the plains.

According to the local population, the terraces have always existed where they are today, having been built by their ancestors. Terraces are not built as fixed permanent structures, but are modified continually in order to adjust the height of the walls and to follow the natural contour lines.

Data relative to the amount of manpower required to build a terrace are not available because terraces are not widely built nowadays. Building is carried out by men, and the choice of site falls to the head of the household. Certain people have mastered the building techniques better than others, with older people usually being more adept in this area than the young, which indicates that skills are not being transmitted adequately to the younger generation. The group questioned for this study reported that while no terraces had been abandoned, no new ones were being built, and they were unanimous that mountainside cultivation would be impossible without them.

The *guimelther* (literally 'house of taro') involves placing a wall of dry stones around the plot of land to be irrigated. The height of the wall depends on the flow-rate of the stream to be used. The wall must be able to protect the *taro* when the stream is in spate. Without the dry-stone walls, the *taro* and the seed-bed would simply be carried away. A channel is constructed upstream of the plot and an exit channel is made downstream. The plot is then divided into compartments using small stone dykes (see the figure below). The development and maintenance of the *guimelther* is similar to the *medoedoe*. Research suggests that the use of this technique has expanded widely in the area over the last few years.

The *guimnda* (literally, 'house of couch grass') consists of a bed of *Cyperus esculentus* which is raised above the level of the soil. The root of this grass is dried and then eaten, as are groundnuts, either whole, or pounded into paste. The bed on which the grass is sown is partially surrounded by a peripheral drain and covered with a series of small drains (see the figure above and below). These

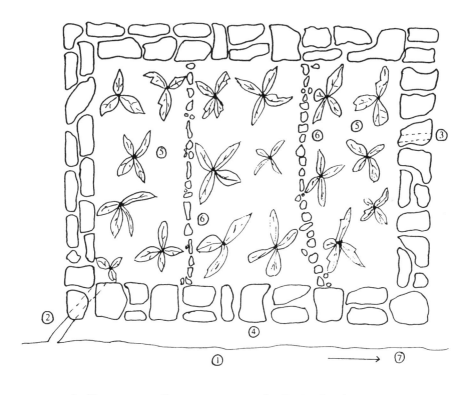

1	Mayo or water flow	5	Garden (taro)
2	Water inlet	6	Stone line
3	Water outlet	7	Direction of slope
4	Low dry stone wall		

Aerial view of guimelther

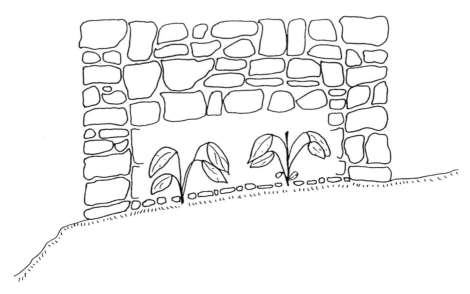

Cross-section of guimelther

drains prevent prolonged contact between the plant rhizomes and the water. The surface area of the *guimnda* is generally between 35 and 100m². To this structure, a mulch of thorny branches is usually added which limits the damage caused by birds and rodents. The choice of site for the *guimnda*, its development and maintenance, depend on women. Older women are more conversant with the technique than younger ones. Local people acknowledge that couch grass yields would be low or nil without this technique.

Ridging (*gid-dankil*) is practised particularly in the valley, the mountain shelf, on the plateaux and in the hills where the soil is saturated with water. Ridges (50cm high and 50cm wide) are built in the direction of the slope, not more than 25m in length. Inside the ridges are weeds cleared during soil preparation. This technique is used exclusively for the cultivation of sweet potatoes. Maintenance of the sweet potato crop is carried out by the young, sometimes helped by their parents. Young people seem more adept at this technique than older people, and it is becoming more and more widespread. Small dykes are made from rows of stones topped with weeds collected after hoeing. Where stones are scarce, the rows are simply made up of the plants and then covered with earth. The dykes rarely exceed 25cm in height and are built on gentle slopes. The main function of the dyke is to retain water on the cultivated plot, to reduce rainwater run-off and to stimulate infiltration.

Micro-dams are built over ephemeral streams to reduce the water speed and to increase infiltration which feeds the groundwater and wells. The dams also retain large quantities of water which can be used for crops, livestock and people. These dams are inspired by traditional dams but have their own

Aerial view of guimnda

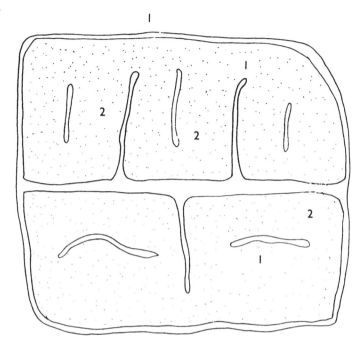

Cross-section of guimnda

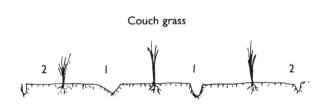

Couch grass

I	Ditches
2	Plots

special features. Modern micro-dams are watertight and do not have a spill-way. With the introduction of cement into the structure the dams can be lar-ger, more efficient and better built. The cost of building such a dam is 300,000–600,000 Fr CFA, depending on the quality of the material used and the dimensions.

Biological techniques

Soil enrichment

Apart from chemical fertilizers, whose use is constrained owing to its prohibitive cost and limited availability, local people employ several methods to maintain or improve soil fertility in their fields.

Poultry is usually kept in the huts overnight, so droppings are easily collected and stored for use as a fertilizer, especially in kitchen gardens. Because of its organic content, the residue of millet-threshing is spread around the stems of crops to improve soil fertility, while ash from the fire is gathered up and applied to the fields.

Soil preparation takes place at the end of the dry season, during which time crop and vegetable residue is burnt, and ash is scattered on the fields. At the start of the growing season, small stock are kept indoors, and their manure is collected and spread on the fields.

Generally speaking, the cultivated plots beside the huts are the first to be fertilized. These plots contain various vegetables, corn, tubers and taro, as well as millet or sorghum. More distant plots are fertilized, but only when compost is available. The head of the household is in charge of collecting the manure and choosing the site for its application. The transport and spreading of manure is done in early May by men, women and children.

Mulching of weeds and leaves is practised exclusively by women, particularly in the *guimnda* fields after sowing, but also in other fields after the second hoeing. Its effectiveness is increasingly acknowledged by local people and its use is on the increase.

Litter from different trees and bushes left in the fields constitutes an important source of organic matter. The leaves of species such as *Acacia albida* are distributed over the fields, as are rotten leaves and millet stalks. Weeds are left on the soil surface and incorporated when the soil is earthed up.

Biennial sorghum/pearl millet rotation

Sorghum is a deep-rooted plant, while pearl millet is shallow-rooted. Biennial rotation of the two crops allows the deep soil and surface soil to restore themselves alternately. Some farmers also recognize that rotation breaks the reproductive cycle of the adventitious plants such as *striga* and parasites like caterpillar which attack sorghum.

The role of agroforestry

In the Mandara Mountains trees, shrubs and grasses are put to many uses, such as the strengthening of terraces, the fertilization of fields, controlling erosion on the terraces, stream banks, around dwellings and along tracks, and the demarcation of cultivated plots.

Terraces are by far the most widespread physical structure, and are frequently complemented by biological techniques. For example, terraces receive organic manure, while simultaneously supporting trees.

The techniques encountered tend to be multi-purpose: the terrace reduces soil loss and run-off and encourages the infiltration of water into the soil, while mulching on the *guimnda* protects the couch grass seeds from predators and combats erosion, evaporation and run-off.

All the techniques are carried out by the family work-force. The household head is the decision-maker when it comes to choosing a technique and locating the site for it. Formerly, there was a clear division of male and female tasks, with the building of terraces and the repair of walls the exclusive preserve of men. Nowadays, it is the women who maintain the terraces, rarely assisted by the men, and all work is done manually.

Projects and organizations operating in the Mandara Mountains

Various organizations are working in SWC in the extreme north. They include the Ministries of Agriculture and Livestock. The parastatal organization SODECOTON is by far the most significant organization in the area, owing to its influence on cotton production. The Soil and Water Conservation Project of the Mandara Mountains (Conservation des eaux et des sols Monts Mandara has initiated a number of developments such as the construction of modern micro-dams, the protection of stream banks, the planting of trees and the construction of improved terraces. Several NGOs are also working in the area on a variety of issues, such as agriculture and reforestation, and the construction of wells and micro-dams. Their activities, however, have been concentrated on the plateaux and surrounding plains. Mountain communities have hardly been reached; indeed, there have been initiatives to encourage mountain dwellers to move down to the plains.

CONCLUSION

The Mandara Mountains are characterized by an unfavourable climate and difficult soil conditions. Nevertheless, a mountain civilization has managed over the centuries to develop an intensive production system. The defining feature of this system is the building and upkeep of structures for the conservation of soils and water in the form of terraces, stone walls, small dykes, irrigation and drainage channels. These structures are supplemented by a range of biological measures for the restoration and maintenance of soil fertility, which include agroforestry, composting, mulching, and the use of combination crops and rotation.

25

NEW PERSPECTIVES ON LOCAL CONSERVATION TECHNIQUES

A case study from Malawi

Julius H Mangisoni and G S Phiri

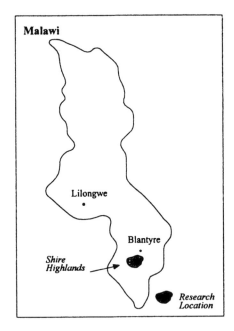

Land degradation is a serious problem in Malawi. Crop yields are steadily declining owing to poor land husbandry practices, and numerous efforts to implement conservation measures have not succeeded in reversing this trend. Many of these measures, often mechanical, have been alien to traditional concepts and practices. As local techniques have been neglected by government agencies, this chapter attempts to unravel indigenous soil and water conservation (ISWC) practices which could be developed and thus contribute to the continued sustainability of smallholder livelihoods in Malawi.

INTRODUCTION TO THE REGION AND PEOPLE

Malawi is a small, land-locked country in southern Africa, bordered by Zambia in the west, Tanzania in the north and north-east, and Mozambique in the east, south and south-west. Current estimates put the population of Malawi, which is growing rapidly at 3.7 per cent per annum, at 10 million. The overall population density in Malawi is estimated at 85 persons per km^2 (National Statistical Office, 1994).

This study was carried out in Mulanje and Blantyre Shire Highlands Rural Development Projects (RDPs) of the Blantyre Agricultural Development Division (ADD) in southern Malawi. Blantyre ADD is about 2000m above sea-level and has a relatively high population density, ranging from 220 persons per km^2 in Mulanje to 292 persons per km^2 in Blantyre District (Malawi Government, 1993). This study relied mainly on data collected through a household survey of 75 smallholder farmers, grouped into four categories according to land-holding size. Both areas are mostly inhabited by the Lomwe, Yao, Chewa, Sena and Mang'anja ethnic groups.

Blantyre has a tropical continental climate with four seasons (Nzima, 1985):

1. Hot and dry (September to early November).
2. Hot and wet (mid-November to March).
3. Cool and moist (April to May).
4. Cool and dry from June to August.

The temperatures in the area vary from 26 to 28°C in October and November to less than 15°C in June and July. The main rains occur between December and April, with the early rains coming around October and November. There is considerable variation in the rainfall, which ranges from 500mm in Blantyre to about 1300mm in Mulanje RDP. Most of the rain falls in the form of thunderstorms and local showers.

In terms of soils, Blantyre RDP is dominated by latosols, with some vertisols in the low-lying *dambos*. Areas with slopes of greater than 12 per cent mostly have lithosols which are generally poor for crop production. Sandy loams (*makande*), sandy soils (*mchenga*) and clays (*katondo*) are also common in the area. In Mulanje RDP sandy clays and clays (*makande*), as well as red soils (*katondo*) are common. These soils are of low-to-medium fertility and are generally shallow.

Using and owning land

Land ownership in the region falls into three main categories: public land and forestry reserves which are owned by the government, leasehold or freehold land which is in private hands, and customary land tenure in which the land belongs to the community and not to individuals.

The chief is the only person who is empowered by tradition to distribute the land under customary tenure. Insecure land use is not an issue in Malawi because, although smallholder farmers cultivate customary land, they have a considerable amount of security of land use. However, over the years the amount of land under customary tenure has shrunk owing to the expansion of large cash-crop estates in the region.

Smallholder farms consist of scattered plots, usually less than 1ha in size. The majority of households have more than two plots to enable them to have at least one fertile piece of land. Population pressure and increased encroachment on to marginal lands has increased the number of fragmented farm plots in the region.

The main crops grown by smallholder farmers are maize, groundnuts, tobacco, sweet potatoes, cassava and beans. Maize is the main food crop grown by all households, while cassava and sweet potatoes are also grown as food and sources of income, together with tobacco, beans and groundnuts. Other crops, such as bananas, citrus fruits, chick peas, ground beans, avocados and vegetables, are grown for both cash and subsistence needs.

There are marked differences between cash-crop estates and smallholder landholdings. There is serious erosion on over half of the area occupied by smallholder farms, while the tea estates practise a number of externally introduced conservation measures, including stormwater drains, waterways, contour bunds and vegetative cover – eg, tea bushes and Katomboro grass planted in places where tea bushes have failed. Blantyre has about 42,000ha of tea estates, over 90 per cent of which are owned and managed by foreigners. Thus, while tea is grown on the best agricultural land, smallholder farmers are relegated to marginal areas which require even greater investments in SWC techniques, investments which are often lacking owing to land and labour shortages.

Current pressures on smallholder farming practices

With current population growth rates, it is estimated that the amount of land per capita will decline from 0.46ha to 0.29ha (particularly on customary land) by the year 2000. The shortage of arable land is already forcing smallholders to cultivate marginal land which is more prone to erosion, such as very steep slopes and protected catchments. This has also resulted in extensive deforestation, which is currently estimated at 3.5 per cent per annum (Malawi Government, 1992).

Shortage of labour
There is significant seasonal and permanent out-migration to urban centres from the Blantyre Agricultural Development Division where the long dry season forces young men and women to look for employment from a few months to a

couple of years before they return to the village. Any money earned is frequently invested in agricultural activities such as small livestock (goats, sheep and poultry) and non-agricultural activities such as beer brewing, tailoring, fish-mongering and trade.

The labour shortage owing to the high rate of out-migration is particularly serious for female-headed households, which account for about 30 per cent of all smallholders. The female head of the house has to bear the full weight of both household and agricultural work: gathering firewood, collecting water, pounding maize and doing other household tasks, all of which reduce the amount of labour and time that is available for productive agricultural work.

Low-yielding crop varieties

The continued use of low-yielding unimproved varieties of maize, together with the limited use of fertilizer, has further reduced the fertility of the soil. The adoption of high-yielding varieties is still low, and only about 25–30 per cent of smallholder farmers use fertilizers. Monocropping is also practised on most farms and, without the addition of organic fertilizers, yields are as low as 400–800kg of grain per ha, depending on the soil type.

In the absence of inorganic fertilizers, leguminous crops and animal manure are the most important sources of additional nitrogen. Farmers keep a variety of livestock, such as cattle, goats, sheep, pigs and poultry, although poultry is the most important species in smallholder agriculture. Livestock could potentially play a very important role in providing fertilizer for crops, but less than 10 per cent of farmers use manure, compost or mulching as fertilizer. Livestock populations are currently too small to provide adequate manure for fertilizer and the Ministry of Agriculture has done very little to encourage farmers to make compost for use on their fields.

Given these constraints and the current high level of soil erosion on land occupied by smallholder farms, it is important to consider how current indigenous SWC practices could further contribute towards the long-term sustainability of farming in the region.

THE ROLE OF INDIGENOUS SWC PRACTICES IN THE SMALLHOLDER FARMING SYSTEM

Farmers experiment with a wide variety of locally adapted SWC techniques which are described below.

Mulching and vegetation barriers

Mulching and vegetation barriers were probably introduced in the 1950s and are common in low-lying areas known as *dambos*. The barriers are normally found at the edge of a small gully on top of a field within the *dambo* area. A thick line of sugarcane is reinforced with a strong ridge of crop residues from maize, sugarcane, bananas and millet. In most cases the barriers are constructed by the farmers using family labour and the traditional hoe. The estimated average dimensions are 30m long, 2m wide and 1m high (see the figure below). The vertical spacing ranges from 5m to over 20m, and the technique seems to be quite effective in controlling soil erosion.

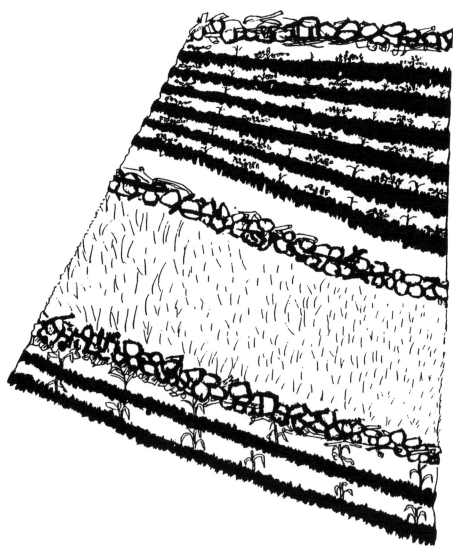

Stone bunds and mulching using tree branches

Banana contour strips

The Ministry of Agriculture recommends the use of contour vegetation strips made of vetiver or napier grass. Vetiver is widely adopted because it is deep-rooted, perennial and unpalatable to livestock, while napier grass is palatable to livestock and therefore susceptible to damage.

Farmers have modified this technique by replacing vetiver or napier grass with bananas on the contours which also produce food for the family. These are planted very close together with about 1–2m between plants and the strips are planted at 10–20m intervals. The technique is not altogether successful as the bananas are not usually planted in infiltration pits (which could further reduce run-off). More recently, farmers have begun to plant fruit trees, such as citrus, pawpaw, pear and leguminous shrubs, in some places to reinforce the strips. Fruit trees are an important source of firewood and food for the family, and the trees further protect the fields from soil erosion. In all, the system potentially provides an effective barrier to run-off, trapping soil and water from upland.

Border planting of sugarcane or elephant grass

Another common conservation technique (particularly in the *dambos*) is planting sugarcane and elephant grass along the borders of farms, roads and homesteads, or scattering it around the homestead. Sugarcane or elephant grass is planted on ridges made around the fields and within fields. Sugarcane can be used for food or it can be sold to provide for other family needs, and elephant grass is used as fodder for livestock or for construction when it is left to mature. However, a disadvantage of border planting is that both sugarcane and elephant grass potentially compete with crops grown in the field.

Tree planting

Tree planting was introduced in the area by a Forestry Department Programme in the 1970s which encouraged farmers to have their own woodlots. Farmers have since modified the recommendation to suit their needs for soil conservation, firewood and poles.

Trees are planted at the foot of a hill and around fields to protect soil from erosion. Eucalyptus is the most commonly used species because it makes good firewood and poles. Trees are closely spaced at 1.6m between rows and 1.6m between plants. Eucalyptus itself is not effective in stopping erosion as undergrowth often develops poorly under this species. Farmers thus slash an under-layer of grass two to three months into the rainy season to form an effective mulch that reduces erosion.

Cassava/groundnuts-reinforced tied ridges

Using hand-hoes, farmers make tied ridges at about 6m intervals to control run-off and to conserve moisture. Many have planted cassava or groundnuts on the ridges in order to reinforce them. This technique appears to be very effective on gentle slopes (up to a 5 per cent slope) because any excess water is safely discharged into the next furrow with minimal damage to the ridges.

Stone and sand sack lines

Old fertilizer sacks filled with sand and small stones or just a line of big stones across a gully are constructed to stop the process of gully erosion (see figure on next page). In some places, slopes are systematically treated with stone lines in order to promote the gradual formation of terraces through the deposition of sediments. The stone lines are set at intervals of 10–30m and are used on any slope where stones are readily available.

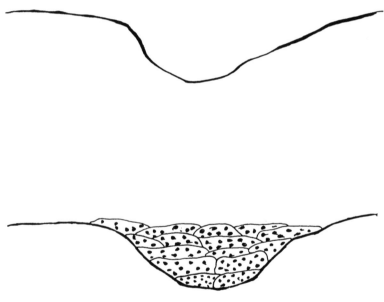

Stone/sand sack lines across a gully

Planting bananas and sugarcane in gullies

Farmers consider gullies a serious threat to their land and only plant bananas or sugarcane as a way of controlling or reclaiming them. In some instances, these have been quite effective, as evidenced by the apparent filling up of the gullies, but in other places the mere planting of bananas and sugarcane has forced the

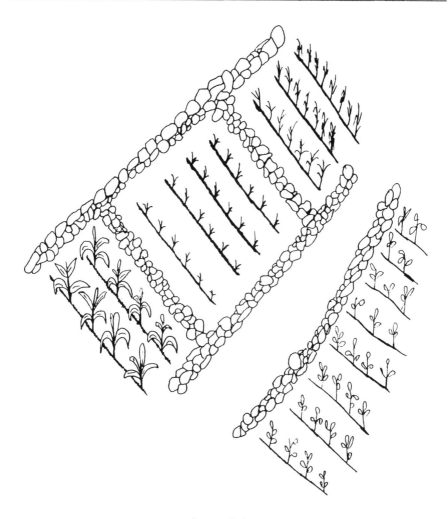

Stone lining

water to change its course, thus creating other gullies. The silted portions of the gullies are used for planting arable crops, such as beans, maize and sweet potatoes.

Silt traps or check dams in gullies

Silt traps in gullies are used by farmers when they want to rehabilitate a degraded piece of land. To prepare the silt traps, poles are erected in the soil, forming a line across the gully. Later, bamboos are either nailed or tied on to each pole, thereby forming a tight mesh that can easily trap soil from upland, which eventually assists in filling up the gullies (see figure on next page).

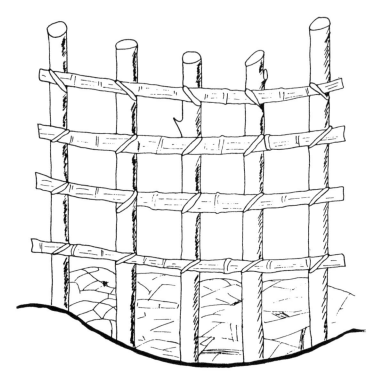

Silt trap in gullies

The dimensions of the traps depend on the size of the gully, but in most cases they would be no less than 2m long and spaced at a distance of 10–15m. Silt traps are restricted to the gullies and may be constructed over their entire length. It appears that the technique evolved from within the area and has spread to other parts of Blantyre, where gullies are more pronounced.

Combining different methods

All smallholder farmers practise indigenous SWC techniques to some extent, in most cases in combination with other methods (see the figure opposite). For example, contour banana strips are frequently used together with tied ridges planted with cassava or groundnuts; stone lines and silt traps are used with the planting of bananas or sugarcane in gullies. The extent to which these indigenous techniques are adopted and are successful is, however, partly dependent on socio-economic circumstances. While mixed cropping, ridging, napier and banana planting in gullies are practised by nearly all farmers, the wealthier farmers who have larger holdings tend to practise non-traditional techniques such as bunds, as more land and labour is needed for their construction.

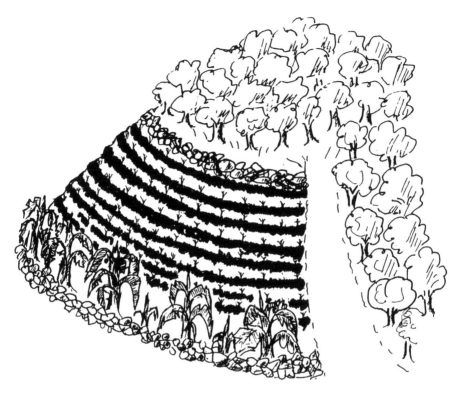

Mulching/stone bunding

Male-headed households have adopted stone bunds, napier grass and banana planting in gullies more than female-headed households, probably because of land and labour requirements. Female-headed households tend to have smaller holdings and more acute labour constraints. Although women take part in conservation techniques, their involvement is mostly in construction and only rarely in maintenance and cultivation.

Government agencies and indigenous conservation techniques

To date, there are no concrete data to show the benefit of local SWC techniques in terms of yield improvements. Soil-conservation experts recognize the existence of such techniques, but there is little evidence that they support the maintenance and expansion of locally adopted measures. The Department of Land Resources and Conservation rarely talks about indigenous SWC techniques, and farmers indicated in the survey that the only indigenous techniques that are being promoted are mulching and tree planting on degraded lands.

During our fieldwork, we have seen no evidence that farmers themselves are making efforts to improve the technical efficiency of these conservation techniques. The only thing that farmers do is to restore the technical efficiency of the existing methods through the constant maintenance of the SWC techniques they deem to be effective.

Efforts are being made to intensify agriculture in Blantyre ADD. Farmers practise early planting in order to gain ground-cover in the early phases of the rainy season, regular weeding and incorporation of crop residues for soil structure improvement. Farmers also practise mixed cropping and, where leguminous crops are involved, there is some nitrogen fixation for the benefit of other crops to be grown on the same field.

CONCLUSION

Modern techniques introduced from elsewhere have not been successful in erosion control for smallholder farmers in Malawi. They require a level of labour investment which farmers cannot afford and they also use up much of the farmers' land which is desperately needed for crops. There is, therefore, a need to go back to the farmers and to focus attention on the techniques that farmers are already using with the aim of jointly improving their efficiency. Research and support should now focus on indigenous SWC practices. To date, there are no data in Malawi that can be used to explain the benefits of various indigenous practices. However, this study demonstrates that it may be easier to improve traditional SWC practices which farmers already know about than to introduce modern techniques that are currently advocated by the government, such as graded bunds, terraces and marker ridges, which in many cases are incompatible with the smallholder farmer situation in Malawi.

The common denominator of indigenous SWC techniques is that, apart from using local materials, they generate food for people or fodder for livestock. In a land-scarce area like the Blantyre District, farmers want techniques that produce more food, firewood, poles and fodder for livestock, while at the same time helping to conserve soil and water. Techniques that consume too much land and do not bring these additional benefits have very little future in the area.

26

THE RAPID EVOLUTION OF SMALL-BASIN IRRIGATION ON THE JOS PLATEAU, NIGERIA[1]

Kevin Phillips-Howard

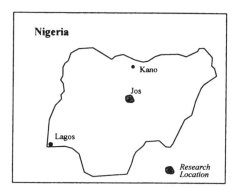

This chapter focuses on one soil and water conservation (SWC) technique – small-basin irrigation – and its rapid evolution on the Jos Plateau in Northern Nigeria, in response to recent social and economic changes. Small-basin irrigation has created new and profitable opportunities for agricultural production. It enables crops to be produced during the dry season as well as during periods of drought in the wet season. Although irrigation has been an important agricultural strategy for many years, the technology and its application have changed dramatically. It is argued here that while the potential to expand small-basin agriculture on the plateau is great, the predominant obstacles remain the substantial investment in labour and capital which are required to exploit it.

[1] Contribution from the Jos Plateau Environmental Resources Development Programme (JPERDP), Jos-Durham University Linkage, supported by the European Development Fund (Project No. 5106.53.41.001) and the University of Jos.

INTRODUCTION TO THE AREA AND PEOPLE

The Jos Plateau comprises a densely populated upland area of 8600km^2 (mean altitude 1200m) which is characterized by cultural diversity, small-scale agriculture and pastoralism. In the early 1970s, population densities ranged from 70 to 139 persons/km^2 in the southern part of the plateau to 140 to 279 persons/km^2 in the northern part. Since then, population densities have continued to increase. In the congested north, intense pressure on land and short fallows have been associated with soils of very poor nutrient status (LRDC, 1978) and sub-subsistence levels of production (Morgan, 1979).

Monthly mean temperatures (20–24°C) are low for the latitude (10°N), while the rainfall (1000–1500mm per annum) is comparatively high. The plateau comprises an undulating plain of basement complex rocks with intrusions of younger granites and basalts forming rocky hills (Morgan, 1979). The plain is cultivable where water is available, but its deep, sandy soils are generally poor and susceptible to erosion. The more fertile soils of flood plains (*fadama*) soils are also cultivable in the dry season, which lasts from October to April.

The much-depleted savannah vegetation that covers most of the plain is made up of grass, shrub and woodland. The 'economic trees' on the farmland and plantations consist mainly of eucalyptus. About 53 per cent of the Jos Plateau is actively cultivated. Of the remainder, about 37 per cent consists of woodland, 6 per cent of shrub/grassland, and 4 per cent of shrubland.

The local farming system was characterized in the 19th century by mixed agriculture, including shifting cultivation and intensive terrace farming in the hills (where many of the settlements were located) and shifting cultivation on the plains. This involved the cultivation of *acca* or hungry rice (*Digitaria exilis*), *dauro* or dwarf millet (*Pennisetum americanum*), *tamba* or finger millet (*Eleusine coracana*) and sweet potato (*Ipomoea batatas*), among other crops. In addition, there was the nomadic and semi-sedentary herding of cattle, goats and sheep. In the colonial era, many settlements relocated on the plain and, according to Adepetu (1986), the traditional subsistence-oriented agriculture of the area was transformed by the development of tin-mining and by rapid population growth. This led to an expansion of the area under cultivation as well as more farmers, and a greater variety of crops was grown as production became more intensive and market-oriented.

Recent socio-economic and rural development trends on the Jos Plateau have reinforced this expansion. The growth and decline of large-scale tin mining have left 316km^2 of mine-spoil, including heaps, mining residue and mine-ponds, along with many retrenched miners. Meanwhile, under the influence of a World Bank-inspired Structural Adjustment Programme (SAP) in Nigeria, there has been a reduced reliance on imports and increased domestic production. A dramatic drop in real earnings in the cities has resulted in improved terms of trade for rural areas and increased employment and rapid market expansion in the agricultural sector. High demand and price increases, partially associated with SAP, have further stimulated market-oriented production, especially of maize and vegetables. These events, combined with the decline in cattle herding as a livelihood strategy, have marked the shift towards small-scale agriculture.

THE EVOLUTION OF SMALL-BASIN IRRIGATION

Irrigated agriculture, or dry-season farming, is an ancient and widespread practice in northern Nigeria which is traditionally carried out on the flood-plains or *fadamas*. This so-called '*fadama* farming' involved small-scale pro-duction of vegetables and other crops during the dry-season. It was traditionally undertaken in small basins adjacent to perennial streams or rivers using a *shadoof* (a counter-weighted lever and scoop). Water was conveyed to more distant basins via earth and stone ramps called *dokki*. The technique was introduced to the Jos Plateau by Hausas from further north, where this practice was thought to have originated (Grove, 1961). Many Hausas came to the plateau to work in the tin-mines and brought with them the knowledge and skills associated with *fadama* farming.

By the 1960s, local plateau people (Berom and others) had begun to use the technique and a government irrigation project was begun near Riyom (Cox, 1965). Up to the early 1970s, however, this activity was limited to a few small areas of *fadama* land (Hill and Rackham, 1973). By the early 1980s, as local circumstances changed, irrigation expanded rapidly on the Jos Plateau. According to Morgan (1985), this was attributable to, *inter alia*, the growing market for vegetables, the introduction of petrol-driven pump-engines and the use of artificial fertilizers. Plenty of labour was available owing to agricultural 'slack' in the dry season and the decline of tin-mining (Adepetu, 1985).

Small-basin irrigation on the Jos Plateau is most commonly found on farms averaging about 0.5ha in size, but ranging from as small as 0.1–2ha. The basins are typically square or rectangular, measuring between 8 and 12m^2 in area. Basin depth varies from about 20–40cm, but each basin is constructed with a more or less level bottom to ensure even distribution of water. The basins are individually flooded using a hoe to direct water from wide channels which measure about 0.5m wide. These are constructed between rows of basins and are fed by gravity with water pumped to a high point on the farm. Pumping is done from nearby streams, lakes or mine-ponds using a 2–3in hose and a petrol motor. The mine-ponds are fed by groundwater and fluctuate in level from as high as 1m below the ground-level at the end of the rainy season (September/October) to 5m or more below at the end of the dry season (May/June) (see the figure on following page).

Construction for small-basin irrigation begins with land clearance, which involves cutting away the bush, digging up grass and roots, extracting stones, breaking up subsoil and burning dead growth. This is followed by thorough excavation of the water channels and basins, the removal of stones and the hoeing of the soil into a fine tilth. The layout of the irrigation system includes the basins into which water is to be pumped, the construction of small terraces

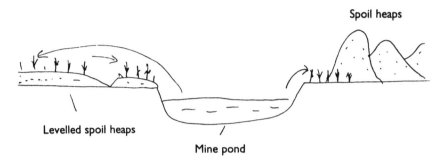

Irrigation on mineland

(either of earth or stone) where necessary, and a network of water channels. On each level or terrace, a series of perpendicular, intersecting ridges about 20cm high are formed. This produces an array of basins separated in rows by water channels (see the figure below).

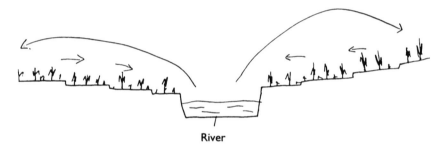

Irrigation with terraces

Maintenance may involve reconstruction of damaged basins using a hoe, although this is rarely needed because of skilled work, and the clearing of weeds from the water channels. However, maintenance requirements are small because the basins and most of the channels are rebuilt annually, early in the dry season (from October).

Among the Hausas, who are predominantly Muslim, all work concerned with irrigation is carried out by men. On the Jos Plateau, many of the Hausa farmers are either retrenched miners or seasonal migrant labourers. The arrangement among the indigenous populations of the plateau such as the Berom is different, with both men and women involved, the latter providing most of the labour.

Arrival of pumps

The replacement of the *shadoof* by the pump-engine occurred rapidly. Among farmers along the Delimi river, the proportion using engines increased from 43 per cent in 1982 to 84 per cent in 1990 (Phillips-Howard *et al*, 1990). The adoption of pump-engines modified small-basin irrigation in several ways. First, the greater flow rates necessitated the construction of higher and stronger basin ridges and the reinforcement of channels; secondly, the expansion of irrigation on to more elevated land became possible; and thirdly, with the collapse of tin-mining, it became possible to exploit the nearly 1000 mine-ponds as water sources.

The expansion of irrigation has been further facilitated by pumping water up to high points, or by the construction of *dokki*, from which the water flows by gravity to distant sites. Relay-pumping is another technique whereby water is pumped into a distant pit from where it is pumped again to higher land (see the figure below). Otherwise, inaccessible areas can be reached by the excavation of channels, the removal of mounds and the construction of elevated hose or piping systems. Where enough capital is available, the simultaneous use of two or more engines makes it possible to irrigate larger and more distant areas.

The incorporation of pump-engines has necessitated various adjustments, including a greater input of labour and faster work. In particular, farmers have had to learn how to manage the comparatively rapid flow of water in order to avoid crop damage and basin overflow. This has involved learning how to adjust the flow rate of the engine and acquiring increased knowledge of the variations in water requirements. Nevertheless, these changes in small-basin irrigation have enabled farmers to use this technique to reclaim substantial areas of abandoned mineland on the Jos Plateau.

Irrigation basins are constructed on mined and unmined land, both of which are generally infertile. Although mine spoil may require more manure, what is more important is the availability of water (rivers and mine ponds), as well as access to fertilizer and urban refuse which is applied as ash. Both types of land are therefore used more intensively the closer they are to Jos and other towns, but are used less so the further away they are from civilization. However, this difference does not remain so important owing to the improvement in access roads and the growing use of a variety of fertilizers among farmers.

Most of the estimated 47,000ha of *fadama* land on the Jos Plateau is subject to some dry-season irrigation except, perhaps, that located far from settlements or tarred roads. The irrigable mineland amounts to about 9400ha. Little of such land is currently irrigated, except that which is close to streams or mine-ponds. The estimated cost of irrigation on these categories of land in 1990 was equivalent to $200/ha/year (Phillips-Howard and Schoeneich, 1992). The major barrier to expansion is the lack of starting capital to buy or hire a pump-engine

Pit   Dokki   Pit

Relay pumping

(costing $300 equivalent in 1992) and to pay for labour and inputs such as fertilizer and seeds.

The irrigable upland is potentially the greatest area for the expansion of irrigation, which is currently limited by higher costs involved in pumping. Altogether, there is about 629,800ha of irrigable land on the Jos Plateau and sufficient water to irrigate almost all of it. However, the local shortage in labour probably means that only a small portion, probably around 15 per cent, could be irrigated.

CONCLUSION

The small-basin irrigation technique enables farmers on the Jos Plateau to generate a large amount of income in the 'slack' period for rain-fed cultivation. The technique is suitable for the production of a wide range of vegetable crops (including tomatoes, carrots, cabbages, lettuce, beans, beetroot, onions and leeks), as well as others, such as sugar cane, wheat, maize and barley. As such, it has been incorporated in some locations into a highly productive and profitable round-the-year system of farming.

With the rapid growth in the market for temperate vegetables, irrigated agriculture should continue to expand into areas where it was not previously feasible. Such areas include elevated land (even on slopes exceeding 5°) adjacent to *fadamas* and the steep-banked rivers and streams and other suitable areas where the creation or exploitation of new water sources, such as bore-wells, become feasible.

The main disadvantage of small-basin irrigation under current circumstances is that it is highly labour intensive. This is problematic in terms of the amount of labour input needed per person and the increasing cost of such labour. Also, the supply of suitably skilled labour on the Jos Plateau is insufficient – hence, the need for continued reliance on seasonal migrant labour from further north. There is intense competition among farmers and they tend to 'hoard' their knowledge, only training relatives and labourers. This is particularly evident among the Hausas who introduced the technology. Hence the plateau people were excluded and had to learn by watching Hausa farmers in their area. So ethnicity was and remains a barrier to a certain extent. More recently, local people have increasingly adopted the technology as its profitability has become more evident. Along with this there has been a tendency to try to remove the Hausas and to regain land allocated to them.

If the demand for irrigated produce continues to exceed supply and the profitability of dry-season farming continues to increase, it is likely that more and more land will be brought under irrigation. However, this will only be possible if problems related to the recruitment and training of labour, as well as the mobilization of sufficient start-up capital, can be overcome.

27

A SOIL AND WATER CONSERVATION SYSTEM UNDER THREAT

A visit to Maku, Nigeria

E M Igbokwe

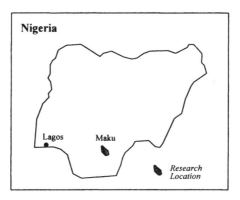

In Maku, an agricultural community on the southern tip of the Udi-Nsukka hill complex of eastern Nigeria, hill farmers have developed an indigenous soil and water conservation system which consists of physical (stone-walled terraces) and agronomic (ridging across the slope with ties or basin listing) measures (Floyd, 1969). There are now signs of decay, abandonment and neglect in this terrace system, which is known locally as *ishi-mgboko*. This chapter looks at the factors and processes that have led to the gradual disintegration of this once sustainable system.

INTRODUCTION TO THE STUDY AREA AND PEOPLE

Maku is located 50km south of Enugu in the Awgu Local Government area of Enugu state. It lies within one of the most populous areas of Nigeria where

---

average rural population density is 335 persons per km^2 (National Population Commission, 1992). Maku's population was 8710 in 1963 and this was expected to double by 1994 (World Bank, 1993). Ifite, Otokwu and Ezioha are the earliest settled villages and occupy the highest land, and Enugwu, Enugwu Eham, Ibite and Eziama are more recently settled villages.

Falling within the northern rain-forest zone, Maku has four dry months, from November to February, with the driest having less than 29mm of rainfall. Annual rainfall ranges from 1600 to 2000mm. Heavy thunderstorms occur at both the beginning (March–April) and the end (September–October) of the rainy season. Hours of daylight are long, although the actual amount of sunlight received depends on cloud cover, rainfall and harmattan haze caused by dust in the dry season (Unamma *et al*, 1985). Annual mean temperatures are high: 25°C in the hills and up to 30°C in the valley bottoms and lowlands.

The vegetation is a derived savannah. However, trees and shrubs are still dense in valleys and sacred groves. On the hill slopes, shrubs predominate and are interspersed by wild oil-palm trees. Soils are formed on resistant sandy shale and siltstone (Floyd, 1969; Igwe, personal communication). Local soil names are based on toposequence. Soils on higher steep slopes are called *ugwu ejirija*, which refers to their gravelly and shallow nature, while soils on the lower slopes and foot-hills are called *ani-nri*, meaning 'rich land'. These lower-lying soils are enriched by deposition from higher slopes and are less prone to natural degradation. The poor sandy lowlands are referred to as *uvom*.

Land-holdings and land tenure

Village settlements are mostly located in palm groves on flat hilltops. People farm the hilltops and the steep slopes close to their houses, but may also have distant farm plots located on lower, gentle slopes with gradients of less than 10°.

The land-tenure system has evolved from communal control to a system of individual (male) ownership. Plots ranging in size from 0.01ha to 0.13ha are owned at various points up and down the slopes, while plots of up to 0.3ha are common in lowland areas. The size of the farm plots tends to decrease over generations; large family size (average household size is six persons) and poly-gamous marriages tend to increase fragmentation (Lal, 1993). The average number of plots per family is currently six. Most local people are fully aware of the fact that fragmentation can lead to uneconomic holdings, so allocative processes now favour consolidation where possible. Families with less land have an option to rent on an annual basis or to share-crop. In addition, long-term leases for up to ten years are common between farmers and temporary migrants.

Farming systems in the Maku hills

Compound farms consist of individual plots where yams, cocoyams, maize and vegetables are grown on terraced slopes closest to the village (see the figure opposite). A system of crop rotation is used to restore nutrient levels, particu-larly for yam cultivation. The rotation system works as follows:

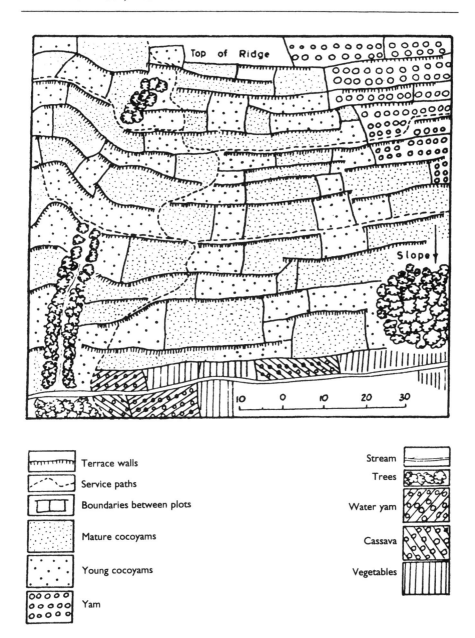

Sketch plan of Maku land use pattern
(from Floyd, 1969)

- Year 1: yams intercropped with maize; chemical fertilizer application
- Year 2: vegetables; heavy organic and inorganic fertilizer application
- Year 3: cocoyam; heavy organic manuring

Farmers do not always start their rotation at the same time and consequently hillsides often have a chequered appearance. Yams are grown on tied ridges constructed across the slope. Land preparation begins in September and is followed by the sowing of early maturing white yam varieties and water yam in October. Chemical fertilizer use, which has gained in popularity in recent years, is usually applied shortly after germination. Certain features of yam cultivation on the terraced slopes are particularly interesting. First, the yam varieties selected are early maturing, provided they are given enough soil fertility maintenance during the rotation course. Yams supplemented with maize therefore provide food during the 'famine months' between June and September when food reserves are depleted. Secondly, farmers are aware of the high level of nutrients required by yams and therefore start organic manuring two years before the plot is sown. However, it is widely believed that organic manures bear pathogens which infest the tubers and, above all, generate excessive plant canopy with poor tuberization, so farmers prefer to use chemical fertilizers during the crop growth period.

Other crops are sown when the rains are regular in March. On the lower slopes and foothills, yams, cassava and cocoyam are grown on tied ridges without any strict rotation. Slopes are gentle, so terraces are not required.

Soil fertility management

On the lower slopes and foothills, bush fallowing is practised to maintain fertility, and during this period regrowth of the shrub, *Acioa barteri* (Igbo: *oka/ahaba*), on the plots helps to stabilize soils and provide sticks for staking yam vines. Organic manure, mainly from sheep, goats and compost (Igbo: *uribi*), but also from poultry, is widely used on the terraced farms. The average stock holding per family is six. Sheep and goats are tethered during the cropping season and stall-fed with palm fronds and grass legume mixtures. Compost is prepared from compound waste, grass-clippings and wood ash. Both types of manure are carried to the farm by women and children during the dry season (November–March) and spread on plots reserved for growing cocoyam and vegetables in the rotation. Palm leaves (Igbo: *nkwukwo*) and leaves collected from *Acioa barteri* fallows on the lowlands are spread on cocoyam plots. The palm leaves are known to be rich in potassium which is essential for tuberization.

Cattle used to roam freely, but with the shift towards intensive cultivation, their destructive effect on crops and terrace walls became evident. Individuals and groups now rear cattle in small herds in paddocks, which makes it easier to collect the dung.

Chemical fertilizer

Because of the intensity of land utilization, the inadequate supply of organic manure, and the cheap price of subsidized fertilizer in Nigeria until recently, chemical fertilizers have been widely adopted by hill farmers in Awgu area (Igbokwe, 1985). However, following implementation by the government of the Structural Adjustment Programme (1985–93), subsidies on fertilizers have been gradually removed, and current prices are beyond the reach of many farmers. It is probable that chemical fertilizer use is on the decline and that of organic manure is increasing, although data which could support this supposition are not available.

Erosion

The major cause of erosion in Maku is high rainfall, although its impact is magnified by the loss of vegetative cover and the growing intensity of cultivation without fallowing. Heavy rainfall also causes rapid loss of plant nutrients through leaching. While sheet erosion is commonplace in Maku, most farmers do not recognize its presence and effects. In contrast, the effects of gully erosion are well known and most efforts in terrace farming are aimed towards preventing this. On well-tended fields, gullies are rare. Erosion of all forms, and especially gully erosion, is widespread on neglected or abandoned plots, and between farm plots on higher slopes.

THE SWC SYSTEM IN MAKU

The SWC system in Maku is mainly confined to the steep slopes, although in some cases agronomic techniques, such as ridging across slopes, construction of earthen bunds and contouring, may also be used on the lower slopes. The origins of SWC in Maku are unclear. One version of oral history maintains that as the transatlantic slave trade intensified, local people erected stone walls and dug-outs at the base of the hills to fortify their villages against slave raids. It is argued that this period of confinement compelled people to find ways to subsist on the rocky soil.

Stone-wall terraces

Terrace walls are built with stones that have been excavated from the slope using metal dibbers, mattocks, hoes and baskets. Farmers select the lowest contour bordering their plot and excavate to a depth of about 30cm for the foundations. This is laid with stones in two or more straight lines. The thickness of the wall may vary from 25cm to 35cm. Construction starts in the early part of the dry season when the soil is very loose. The male head of the household, together with a co-operating neighbour or in-law, takes turns to hoe the gravelly earth facing the hill to fill the base openings. This base is further strengthened by heaping sifted rock fragments. Another layer of stone is laid and the process repeated until a desired height of around 30cm above soil level is reached. Earth is hoed continuously from the slope towards the wall in subsequent years to level the ground. Women and children sieve the soil with baskets to extract gravel and pick up stones. These are piled on the boundaries between plots to create barriers which slow down the speed of run-off. The long-term objective is to reverse the slope of the terrace platform and, aided by tied ridges and deep tillage, water gradually percolates into the soil after each rain. Excess stones are packed in heaps to be used for raising the height of the wall when necessary or for future repair work.

Since plots are small, terraces do not form a continually running bench around a hillside, but are interrupted and re-formed at regular intervals so that they overlap at both ends of the slope. Terraces are usually not cropped in the first year following construction, but are left to consolidate.

Maintenance of terrace walls

Wall maintenance is an on-going process, and is especially important after a heavy rain since any serious break in the wall may lead to the crops being washed away. General repair work is concentrated in the period at the end of the rainy season when there is less work to do on the farm. This laborious work is usually undertaken by the male head of the household, although the assistance of male neighbours and in-laws may be requested. Terrace repairs require investment of eight working days for each 4–5m length of wall.

Advantages and disadvantages of the SWC system

To the typical Maku terrace farmer, the system has the following advantages:

- Soil erosion is checked.
- Water conservation enables farmers to start yam planting as early as October, while those living in outlying lowlands must wait until the rains in March. Above all, it has enabled year-round vegetable cultivation. Vegetables have ready markets in major urban areas such as Enugu and Onitsha, especially during the dry season.
- The system encourages continuous working of the soil from year to year. This provides increased soil depth for yams and encourages dynamic soil-forming processes.

Some of the disadvantages cited are:

- The labour intensity of the system at a time when many families face labour scarcities.
- Regular inspections of terrace walls are time-consuming.
- The system consumes scarce farmland.
- Terrace walls provide homes to rodents which destroy crops, especially tubers. Snakes often inhabit the terrace walls, making reconstruction work hazardous.

The majority of farmers who have adopted these SWC techniques are ordinary villagers who farm mainly for subsistence and who live in the earliest settled villages, Ifite, Otokwu and Ezioha. However, even here, new terraces are no longer being constructed on any significant scale. Terrace farming is far less common in the later settled villages because farmers have access to near-level land on the south-western and eastern slopes and valleys, and this provides land for extensive farming and fallowing.

Trends and constraints to the expansion and maintenance of SWC in Maku

In 1969 Floyd wrote, 'There seems to be no alternative to the fact that a greatly increased rate of emigration from the terraced hill land must be achieved if those who remain are to have any sort of life other than one of constant drudgery and privation.'

The current situation is not as clear-cut as Floyd suggests; well-maintained plots within intact walls are intermixed with abandoned and dilapidated terrace walls. Very few (less than 0.5 per cent) of the existing terrace plots have been constructed in the last ten years, and one out of every ten plots is being abandoned yearly. Abandoned plots can be found in all villages in Maku. Because of the increasing costs of yam production, especially the purchase of seed yams, more and more male farmers are giving up or reducing the scale of yam production and giving more land to cassava and cocoyam production.

The accumulation of massive oil wealth in the early 1970s led to a rapid expansion in education, health, industry and urban development, and the abandonment of agriculture, especially among the educated classes. Rapid population growth meant that existing agricultural technologies failed to produce an adequate subsistence. The outcome is that for many rural dwellers migration in search of non-farm occupations with quick cash returns has become the order of the day. Temporary out-migration from Maku can involve the periodic movement of whole families into urban centres like Enugu and Lagos. Seasonal migration is more common among poorer male farmers who migrate between April and December every year to villages located near the major urban centres. Prevailing beliefs and sentimental attachments to ancestors and land mean that permanent migration is unknown in the area.

Labour scarcities, in combination with the drudgery of the work, are major constraints to the management of terrace agriculture in Maku. The system requires the labour of physically fit young men, but given an average age of 40 for male rural dwellers in Eastern Nigeria (Unamma, 1985), it is doubtful whether the system can be sustained. A second change is the emphasis placed by the government since the late 1970s on achieving universal primary education. This means that children who earlier provided farm labour now attend one of the four primary and two secondary schools in the area. Education has greatly expanded young people's horizons and given rise to strong preferences for non-farm occupations.

Stone quarrying in Maku has drawn some people away from their farms and has led to the pilfering of terrace-wall stones which can be sold for cash. Palm wine tapping is another occupation which has consistently drawn men away from terrace farming (only males are allowed by custom to climb palm trees and tap palm wine). As existing wild palms grow old and because the shallow hill soils in Maku preclude the large-scale development of palm plantations, tappers are increasingly migrating on a seasonal basis to suburban towns.

There is no established mutual work-group system among males in Maku, except those concerned with home- and road-building, which might ameliorate the agricultural labour shortages. Terrace construction is a tedious, slow and time-consuming task which keeps every male member of the village busy in their own plot at the same time. In consequence, little time is left to work in rotation. Moreover, hired labour is generally not available for construction.

The migration of male heads of families or their engagement in non-farm occupations often leaves women responsible for terrace maintenance. However, women's heavy workloads often prevent them from carrying out maintenance and repair work on existing terrace walls, while their lack of familiarity with terrace construction means that new terraces are not being built.

Temporary migration sometimes forces migrants to lease or rent land to other farmers, or to leave land in the care of immediate relations. These arrangements rarely work to the advantage of the migrant since plots are often relentlessly cropped without manuring. Decreasing the supply of organic matter weakens soil structural stability, leading to increased run-off which may cause walls to collapse. Moreover, long-term investment in the land, including terrace maintenance, tends to be neglected because the returns and benefits of this work cannot be assured. In contrast, where long-term leases of up to ten years have been arranged, responsibility for terrace-wall maintenance and fertility management are agreed verbally, and a small token fee per plot is paid to the temporary user by the owner. This arrangement has had a more positive effect on soil conservation than the other shorter-term rental or loan arrangements.

Another major constraint to the sustainability of the system is the dwindling supply of organic materials for manuring. High-population densities mean that large portions of the hill are newly cleared and brought into cultivation. This removes a vital source of vegetative material for compost-making. Fallow periods in the lowlands reclaimed by *Acioa barteri* cultivation are getting shorter, which reduces the supply of leaves. Palm leaves are also in short supply as old plantations are cut down and not replanted. The production of animal manure has also fallen below demand, while the shift towards chemical fertilizers is being undermined by the new high prices for these inputs.

The availability of other production inputs, such as herbicides, insecticides, improved varieties of crops and breeds of animals, can also be problematic. The increased use of fertilizers increases weed incidence, with the result that household labour must be switched away from other management operations. Improved cassava varieties are available and are widespread in Maku. However, indigenous crops such as yam and cocoyam, which are widely grown on the hills, have not undergone any noticeable biological improvement and their yields remain relatively low. The use of insecticides is unknown because of cost and supply problems, despite severe occasional outbreaks of pests and diseases. To date, most farmers have relied on indigenous methods, such as spraying wood ash or the careful selection of planting materials.

Improved varieties of sheep and goats have not been introduced in Maku, so poor returns are unlikely to encourage expansion and hence to provide the essential by-product – manure – which is required for maintaining soil fertility. The main implication is that yields, and hence incomes, will remain low. This gives added stimulus to out-migration. Yet permanent resettlement is considered out of the question by local people. As one farmer argued, 'You cannot transfer land; whoever shreds his mat sleeps on a bare floor.'

Role of government and NGOs

Since 1986 intensive extension contact with selected farmers in Maku has been pursued under the World Bank-assisted Agricultural Development Programme. However, since the innovation packages disseminated do not take into account the ecological specificities of different areas, localized SWC problems in Maku have not received attention. Within this extension system farmers are expected to bring their specific problems to the attention of extension agents who may offer solutions or refer the problems to research units. Yet this approach has not worked very well in Maku because, as in much of West Africa, soil conservation falls outside the remit of many extension workers (Lal, 1993).

CONCLUSION

The Maku SWC system stands out as an example of a sustainable, indigenous agricultural practice which is gradually being overwhelmed by ecological and socio-economic factors. Pressure on the land is resulting in a growing intensity of land use and the rapid loss of vegetation. The expansion of education, increased participation in non-farm occupations and the high incidence of out-migration among the youth, remove the very people who have the energy to maintain and expand the system. One important lesson that can be drawn from this situation is that when a system reaches its capacity, decline may set in unless external interventions are initiated to arrest such a trend. In Maku, no such intervention has so far been developed.

28

EVOLUTION OF TRADITIONAL TECHNIQUES OF SOIL CONSERVATION IN THE BAMILEKE REGION, WEST CAMEROON

Paul Tchawa

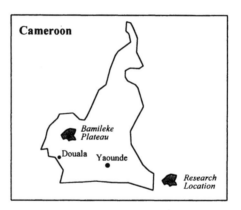

The Bamileke highlands provide a well-watered and fertile environment for a diverse range of crops. Farmers have successfully adapted their farming systems to accommodate new crops and marketing opportunities. Ridging, hedges, fallowing and agroforestry practices are used to improve soil fertility, but also to bring important benefits in terms of soil and water conservation (SWC).

INTRODUCTION TO THE AREA AND PEOPLE

The highlands of West Cameroon rise from rolling plains to the height of 2700m in the Bamboutos Mountains. Rainfall declines as you move from west to east, with the south-western slopes receiving up to 3600mm of rain annually. The vegetation has been widely affected by human habitation, with only a few escarpments still covered by primeval forest which is in the process of being destroyed to make way for fields. Elsewhere the slopes are dissected by hedges, while the lowland is mostly covered by *Raphia* palms.

This region is principally inhabited by the Bamileke people and exhibits great inter-regional variation, both physical and socio-economic. The northern part is mountainous, densely populated and more varied than the southern zone which is lower lying, with a harsher climate and less populated. The northern part is represented here by Bafou and the southern zone by Bakou, the two locations showing a sufficient contrast to encompass the complexities of the region.

The human context

Bakou provides an example of a relatively sparsely populated location suffering from under-exploitation. In 1987 the total population of Bakou was 1061 and, of these, only 154 were aged between 15 and 49 years. By contrast, in 1967 the population was over 2000; thus, in the space of 20 years, half the population had moved away. Average population density is 50 inhabitants per km^2 which varies from 15 to 100 persons/km^2. With falling demographic pressure, the slopes are being recolonized by natural vegetation.

The population of Bafou is thought to have been about 50,000 in 1987. It is estimated that in 1976 the overall density of the locality was 240 inhabitants per km^2. Today, the figure has probably reached 275, but this average conceals considerable variation.

Land ownership, economic activities and land use

Bakou demonstrates a difficult and inflexible system of land ownership, accompanied by numerous disputes over land. In some families (23 per cent of cases), there is great uncertainty over land-holding following the death of the household head, before the new heir to the household is properly installed through traditional ritual. Moreover, 41 per cent of family heads live in town, far from the lands they have inherited. Many problems of ownership, attribution and land development remain unresolved as a result. It is not surprising, therefore, that in 1990 fewer than 22 per cent of Bakou land concessions were being fully exploited.

Arable agriculture is the principal activity in Bakou, and crops include maize, groundnuts, red kidney beans, manioc, macabo (*Xanthosoma* spp), taro (*Colocassia*), yams (*Dioscorea* spp), bananas and sweet potatoes (see photo 31). Here, as elsewhere in the Bamileke region, the land is subdivided into particular areas for each group of crops. Thus, macabo and taro generally occupy the lower third of the slopes, manioc and beans are often found in the middle of the slopes, and yams, groundnuts and sweet potatoes are generally kept to the upper third. Crops are usually grown underneath coffee trees. Bakou's current economic situation is particularly weak, with poor roads and limited access to markets.

In Bafou, as elsewhere, the Bamileke people make a distinction between land held by the household and that belonging to the community as a whole. The latter is made up of grazing land, used by everyone in past times, for pasturing goats. Farmers could also cut a small plot from this land for their own needs. In Bafou, there were very extensive areas of such communal lands, of great interest to people from outside the area who, from the 19th century onwards, have sought to acquire stretches for their own use. Following independence, these lands have been effectively divided up between peasant farmers, formerly nomadic herders, and a state-owned tea company. Many conflicts have broken out between different groups, and each has been developing a strategy by which to encroach on the territory of others, while protecting their own patch. This explains the spread of scrubland in the northern part of Bafou. Here, each time that herders take their animals off on transhumance, farmers take the opportunity to settle on this grazing land, put up barbed wire fences, and plant stands of eucalyptus.

Life in Bafou is geared to agropastoral activities. Food crops include maize, groundnuts, beans, yams, macabo, taro, manioc, plantains and bananas. Arabica coffee represents the main traditional source of income, having been introduced in 1920 and spread rapidly. The expansion of small-scale horticulture began in 1982 with the creation of a project to develop marshy, low-lying areas for gardening. Gradually, the success of this operation has resulted in horticulture spreading beyond the lowlands to take over the fertile slopes of the Bamboutos (see photo 32). Here, irrigation can guarantee two to three harvests a year of potatoes and maize. Yields are high for both crops, and can reach 6–7t/ha each season. Horticultural production is aimed mainly at markets in Yaoundé and Douala, or Libreville in Gabon. For several years farmers have also been exporting green beans to European markets.

The commercialization of horticultural produce is the determining factor behind current patterns of land use. *Raphia* palms are disappearing from the lowlands and being taken over by fruit and vegetables. By contrast, coffee has declined in importance with the recent fall in prices. The pastoral Mbororo, who have been seasonal visitors to the area for generations, have been increasingly marginalized by these changes in land-use patterns, and the Bamileke people seem to be investing more and more in livestock themselves. Pig breeding and poultry farming are being modernized.

TRADITIONAL MEASURES FOUND IN THE REGION

The Bamileke people employ numerous soilconservation techniques, such as the use of green fertilizers and manure, the integration of livestock with agriculture, the application of crop residues to the land, fallowing, hedges and ridging. Almost all these techniques are aimed primarily at improving soil fertility; protection from water erosion is of only secondary importance.

Ridges

Ridges vary in their width, height and direction depending on the topography, soil depth and water flow. The type of crop to be planted on the ridge and land-ownership structure also influence ridge size and shape. Originally, ridging was carried out mainly in the direction of the slope, but there were several variants, depending on relief.

The size of the ridge decreases in areas where the slope is steeper or the soil is shallow. Ridges are generally higher for root plants such as yams than they are for beans and groundnuts which are planted on ridges of average 20–30cm height. Evidence suggests that the length of the ridge seems linked to the land-ownership situation and to patterns of land use up and down the slope. In most cases, the same ridge never runs the whole length of a slope, and its average length is around 10m.

There are two main stages in the construction of ridges. The first step involves the destruction of the ridges from the preceding crop and the thorough removal of weeds or crop residue from the old ridge site into the furrow. The second stage takes place one to two months later and involves hoeing sufficient soil to cover the vegetation that previously was placed in the old furrows. In this way the new ridges are built up where the furrows of the preceding crop used to be.

It is principally women who construct the ridges, while men take part only in exceptional cases, such as when the ridging has to be done on what was fallow land. Young girls from the age of 12 or 13 know how to construct ridges. Adults tend to organize themselves into work groups (*seuk*) of three to six people who work in turn on each other's plots. Based on family ties or friendships, these work groups have existed among the Bamileke for generations. In polygamous families where there is agreement, the wives often form their own group.

Hedges

Hedges provide a means for defining the area owned by the community as a whole, and by households within this larger group. Hedges running across the hillside provide a barrier which keeps the animals within the upper third of the slopes, thereby sheltering crops lower down from livestock damage. The most commonly found species in these hedges are *Dracaena, Markhamia lutea, Polyscia fulva, Vernonia* sp, *Podocarpus milanjenus*. The majority of these vegetative barriers are grown from cuttings and develop relatively quickly into hedges. They reduce the speed of water flow and infiltration and provide significant quantities of organic material to the soil.

Combination of crops

In order to make the best possible use of his plot, the Bamileke farmer combines various crops on the same ridge, each crop being of different height, form and cycle. Mixed planting provides almost permanent ground-cover, so ensuring good protection against erosion. Groundnuts, maize, beans and even macabo are often found growing together, although they are harvested at different periods.

Agroforestry

A further practice which contributes greatly to the relative stability of the slopes is the planting of trees. Bamileke farmers are experienced foresters and numerous trees can be found on their plots. Fotsing (1990) recorded up to 130 stands per ha. Food crops can be cultivated in the shade of coffee trees whose roots also help to secure the soil. At least four different levels can be distinguished within these systems of cultivation. The lowest is represented by sweet potatoes or groundnuts and beans, then comes maize, manioc and coffee bushes. The third is made up mainly of banana trees, and the final, highest level is built from the larger fruit trees. Careful land use from the summit to the base of the slope is combined with this layered pattern of cropping to produce a remarkable stability in soils in this hilly region.

BAKOU: FROM GRADUAL CHANGE TO INNOVATION

Fallow land

In Bakou, not only has the length of fallow decreased considerably, but the people who regulate it have also changed over the past two decades. It is no longer the matriarch who decides on either the length of time or the areas to leave fallow, and as a result there is a certain amount of disorder in the social hierarchy. If fallow land were still regulated by a single person who was perceived as a legitimate local decision-maker, everyone would abide by the rules. But now land is rarely fallowed other than for the cultivation of yams (two to three years fallow after two years under cultivation) and to that of macabo (four years fallow after two to three years farming).

Hedges

Hedges seem never to have been used with the aim of soil conservation, and are themselves in serious decline, as can be seen from aerial photographs. The destruction of hedges began in Bakou with the removal of those that ran parallel to the contours. These hedges were formerly used to ensure that the goats stayed on the upper hillside. Then subsequently the tall hedges surrounding the farm plots have been progressively cut down.

Ridging

Ridging techniques have evolved very little, and continue to be constructed in the direction of the steepest slope. Neither the period of time for preparing the ground nor the tool used for ridging seems to have altered significantly.

Recently, some farmers have been raising small ridges across the furrows which are usually held in place by a plant, such as banana, manioc or yam, put in for this purpose.

Fertilization

Fertilization techniques have changed very little. Less and less animal manure is available because herding, which was its source, has declined greatly. Waste from the local coffee bean extraction process is used to fertilize the coffee trees, and also benefits the food crops growing beneath them. This organic waste is now also being used on plots where only food crops are cultivated.

BAFOU

Fallow land

Fallow land is disappearing completely from Bafou, having been in steep decline for at least the last 20 years. Only 17 per cent of farms leave land fallow and, not only is land no longer left to rest, but there are up to three crops a year in certain horticultural areas.

Hedges

In the old settlement areas in the southern part of Bafou a comparison of aerial photographs from 1964 and 1985 shows little change in the extent of hedgerows. In Baghonto, for example, in 21 years hedgerows have increased in extent from 34m per ha to 41m per ha. This evidence indicates a situation of relative stability and slight improvement. At Feumok further north, on the other hand, there has been a large increase in the extent of scrub-like vegetation over the same period, which is due essentially to the spread of *Eucalyptus saligna*. In the horticultural areas, recent hedges are often comprised of *Cupressus sempervirens* and *Dracaena deisteliana*.

Ridging and fertilization

This technique is undergoing considerable change in terms of orientation, size, construction technique and the implements used. Here, as elsewhere in the Bamileke region, ridges were originally laid out up and down the slope. But since the 1970s, farmers have begun to construct ridges along the contour and today about 70 per cent of the cultivated area at Bafou features this new method. Large ridges, however, are no longer constructed and those seen today measure around 15cm in height with furrows 12–15cm wide constructed for potato and cabbage crops. The technique used for establishing these new ridges has also had to change. First the earth is simply turned over without any preparation and a few weeks later the farmer marks out the furrows on this loosened surface. Then he sows or plants out seedlings at the same time as spreading poultry manure. In the case of seed potatoes, he covers them lightly with earth; for cabbages and carrots, he waits three or four days, then spreads a chemical fertilizer in the furrows. The fertilizer is applied again three or four weeks later and the furrows are closed over. The furrows then become the ridges and vice versa. These changes have led to farmers using a hoe with a blade no more than 10cm in width. This change of technique in ridge construction no longer allows for the traditional practice of ploughing in organic matter and, as a result, agricultural production is relying more and more on manufactured fertilizers rather than on the recycling of biomass.

CONCLUSION

As time passes, the Bamileke region faces new challenges: demographic growth in some places, abandonment of the countryside elsewhere and changes in farming at a time when land-ownership issues are becoming increasingly complex. But there are good reasons for optimism, as most of the traditional conservation techniques offer the real possibility of continued adaptation as opposed to sudden transformation. Past evidence shows that farmers can be immensely adaptable to changing circumstances, as can be seen, for example, by their success in undertaking coffee growing in what was considered to be an agricultural area that was already exploited to the limit. And in Bafou they have become horticultural producers by making good use of the lowlands which had never been cultivated under traditional systems. In both these cases, farmers have simply acted in their own best interests. This suggests that the challenge of increasing agricultural production in future decades while conserving soils will be faced successfully in Bafou, Bakou and elsewhere in the Cameroon Highlands.

BIBLIOGRAPHY

Adepetu, A A (1985) *Farmers and their farms on four fadamas on the Jos Plateau* JPERDP Interim Report no 2, Department of Geography, University of Durham, Durham

— (1986) *Agricultural practices and adjustments in the region* JPERDP Final Report, Chapter 20, pp 342–55, Department of Geography, University of Durham

African National Congress (1994) *The Reconstruction and Development Programme* Umanyano Publications, Johannesburg

Alvord, E (1958) *Development of native agriculture and land tenure in Southern Rhodesia*, unpublished manuscript, University of Zimbabwe

Anderson D (1984) 'Depression, dust bowl, demography and drought: the colonial state and soil conservation in East Africa during the 1930s' *African Affairs* 83, pp 321–43

Anon (1925) 'The dangers of soil erosion and methods of prevention' *Rhodesian Agricultural Journal*, vol 22, pp 533–42

ARPT (Northern Province) (1986) *A Report on the Identification of Zones for Agricultural Research in the Northern Province of Zambia* ARPT, Kasama, Zambia

— (1987) *Informal Survey and Area Reconnaissance Report Zone 1 Kaputa District, Northern Province* ARPT, Kasama, Zambia

Atampugre, N (1993) *Behind the Lines of Stone: The Social Impact of a Soil and Water Project in the Sahel* OXFAM, Oxford

Aubreville, A (1949) *Climats, forêts et désertification de l'Afrique tropicale* Société d'Edition de Geographie Maritime et Coloniale, Paris

Barbier, J C (1988) 'Expansion et limite d'un bocage d'altitude tropicale, le cas du Pays Bamileke au Cameroun', *L'Homme et la montagne tropicale*, 19th SEPANRIT seminar, Yaounde, 1983, Talence, pp 159–71

Barbier, E (1991) *The role of smallholder producer prices in land degradation. The case of Malawi* LEEC Discussion Paper 91-05, IIED, London

Barning, N M and Dambré, J B (1994) *Les styles d'exploitation; une classification des exploitations dans la province du Sanmatenga, Burkina Faso* Etude PEDI/CRPA, Kaya

Barrett, S (1991) 'Optimal soil conservation and the reform of agricultural pricing policies' *Journal of Development Economics*, vol 36, pp 167–87

Beinart, W (1982) *The Political Economy of Pondoland* Cambridge University Press, Cambridge

— (1984) 'Soil erosion, conservationism and ideas about development: a southern African exploration, 1900–1960' *Journal of Southern African Studies*, vol 11, pp 52–83

Bembridge, T J (1987a) 'Some aspects of household diet and family income problems' *South African Medical Journal*, vol 72, pp 425–28

— (1987b) 'Crop farming system constraints in Transkei: implications for research and extension' *Development Southern Africa*, vol 4 (1), pp 67–89

Bishop, J and Allen, J (1989) *The on-site costs of soil erosion in Mali* Environment Department, Working Paper 21, World Bank, Washington, DC

Blaikie, P (1985) *The Political Economy of Soil Erosion in Developing Countries* Longman, London

Bojo, J (1991) *The Economics of Land Degradation: Theory and Application to Lesotho* Stockholm Environment Institute, Stockholm

Bojo, J and Cassells, D (1995) *Land degradation and rehabilitation in Ethiopia: a reassessment* AFTES Working Paper 17, World Bank, Washington, DC

Boserup, E (1965) *The Conditions of Agricultural Growth: The Economics of Agrarian Change Under Population Pressure* Allen & Unwin, London

Boutrais, J (1984) 'Entre nomadisme et sédentarité, les Mbororos à l'ouest du Cameroun' in *Le développement rural en question*, Editions de l'ORSTOM, Paris, pp 225–56

Bromley, D (1992) *Making the Commons Work: Theory, Practice and Policy* Institute of Contemporary Studies, San Francisco

Bromley, D and Cernea, M (1989) *The management of common property resources: some conceptual and operational fallacies* Discussion Paper 57, World Bank, Washington, DC

Bundy, C (1988) *The Rise and Fall of the South African Peasantry* David Philip, Cape Town

Bureau of Statistics (1988) *1988 Population Census Preliminary Report* Ministry of Finance, Economic Affairs and Planning, Dar-es-Salaam

Chambers, R (1983) *Rural Development: Putting the Last First* Longman, London

— (1993) *Challenging the Professions: Frontiers for Rural Development* IT Publications, London

— (1994) 'Foreword' in I Scoones and J Thompson (editors) *Beyond Farmer First: Rural People's Knowledge, Agricultural Research and Extension Practice* IT Publications, London

Chambers, R, Pacey, A and Thrupp, L-A (1989) *Farmer First: Farmer Innovation and Agricultural Research* IT Publications, London

Cheatle, R (1993) 'Next steps towards better land husbandry' in N Hudson and R Cheatle (editors) *Working with Farmers for Better Land Husbandry* IT Publications, London

Cohen, R (1967) *The Kanuri of Bornu* Holt, Rinehart and Winston, New York

Collinson, M P (1972) *The economic characteristics of Sukuma farming systems* Paper No 72, Economic Research Bureau, Dar-es-Salaam

Cox, P G (1965) 'Observations on the agriculture of the Jos Plateau' *The Nigerian Field*, vol 30, pp 84–94

Critchley, W, Reij, C and Wilcocks, T (1994) 'Indigenous soil and water conservation: a review of the state of knowledge and prospects for building on traditions' *Land Degradation and Rehabilitation*, vol 5, pp 293–314

Culwick, A and Culwick, G (1935) *Ubena of the Rivers* Allen & Unwin, London

David, R (1995) *Changing Places? Women, Resource Management and Migration in the Sahel* SOS Sahel, London

Dessalegn, R (1994) 'Neither feast nor famine: prospects for food security', pp 192–208 in A Zegeye and S Pausewang (editors) *Ethiopia in Change. Peasantry, Nationalism and Democracy* British Academic Press, London

Dongmo, J L (1981) 'Le dynamisme bamileke', vol 1, *La maîtrise de l'espace agraire* CEPER, Yaounde, p 424

Ecologist, The (1971) *A Blueprint for Survival* Penguin, Harmondsworth

Ehrlich, P (1968) *The Population Bomb* Ballantine, New York

Elwell, H (1985) *An assessment of soil erosion in Zimbabwe* Zimbabwe Science News, vol 19, pp 27–33

— (1993) *Development and adoption of conservation tillage practices in Zimbabwe. Soil tillage in Africa: needs and challenges* Soils Bulletin 69, Rome

FAO (1983) *Keeping the land alive: soil erosion – its causes and cures* Soils Bulletin 50, FAO, Rome

— (1986) *Ethiopia Highland Reclamation Study* Final Report, Ministry of Agriculture, Addis Ababa

Faulkner, D E (1944) *Soil erosion and the conservation of soil and water in Swaziland* Veterinary and Agricultural Department, Mbabane

Floyd, B (1969) 'Terrace Agriculture in Eastern Nigeria: The Case of Maku' *The Nigerian Geographical Journal*, vol 7 (2), pp 91–108

Fotsing, J M (1990) *Transformations des pratiques pastorales en milieu densement peuplé: Les monts Bamboutos au Pays bamileke (Ouest-cameroun)* Cahier de la Recherche Développement, no 27, Montpellier, pp 32–7

Frost, P, Medina, E, Menaut, J-C, Solbrig, O, Swift,M and Walker, B (1986) 'Responses of savannas to stress and disturbance: a proposal for a collaborative programme of research' *Biology International*, Special Issue 10, IUBS, Paris

Graham, J (1979) *Historical Background to the Wanging'ombe Rural Water Supply Project* BRALUP Research Report No 39 (New Series), University of Dar-es-Salaam

Grove, A T (1961) 'Population densities and agriculture in Northern Nigeria' in K M Barbour and R M Prothero (editors) *Essays on African Population* Routledge and Kegan Paul, London

— (1985) 'Water characteristics of the Chari system and Lake Chad' in Grove, A T (ed) *The Niger and its neighbours: environmental history and hydrobiology, human use and health hazards of major West African rivers* A A Belkaman, Rotterdam

Grove, R (1995) *Green Imperialism: Colonial Expansion, Tropical Island Edens and the Origins of Environmentalism, 1600–1860* Cambridge University Press, Cambridge

Gubbels, P (1994) 'Populist pipe-dream or practical paradigm? Farmer driven research and the projet agroforestier in Burkina Faso', pp 238–43 in I Scoones and J Thompson (editors) *Beyond Farmer First: Rural People's Knowledge, Agricultural Research and Extension Practice* IT Publications, London

Hagmann, Jurgen (1991) *The Soils of Dizi/Illubabor. Their Genesis, Potential and Constraints for Cultivation* Soil Conservation Research Project Research Report 18, Berne

— (1996) 'Mechanical soil conservation with contour ridges: cure for or cause of rill erosion – which alternatives?' in S Twomlow, J Ellis-Jones, J Hagmann and H Loos *Land Degradation and Rehabilitation* Belmont Press, Masvingo, Zimbabwe (in press)

Hailey, L (1957) *An African Survey* Oxford University Press, London

Hawkins Associates (1980) *The Physical and Spatial Basis for Transkei's First Five Year Development Plan* Hawkins Associates, Salisbury

Herweg, K (1992) *Major constraints to effective soil conservation. Experiences in Ethiopia* Paper presented to the seventh ISCO conference, Sydney, Australia

Hill, I D and Rackham, L J (1973) *The Land Resources of Central Nigeria Interim Report on the Landforms, Soils and Vegetation on the Jos Plateau. Landforms and*

Soils Miscellaneous report no 153, vol 1, Land Resources Division, Ministry of Overseas Development, London

Hoben, A (1995) 'Paradigms and politics: the cultural construction of environmental policy in Ethiopia' *World Development*, vol 23, pp 1007–21

Holden, S T (1993) 'Peasant Household Modelling: Farming Systems Evolution and Sustainability in Northern Zambia' *Agricultural Economics*

Hudson, N W (1957) 'Erosion control research' Progress report on experiments at Henderson Research Station, 1953–56. *Rhodesian Agricultural Journal*, 54, 297–323.

— (1987) *Soil and water conservation in semi-arid areas* Soils Bulletin 57, FAO, Rome

— (1991) *A study of the reasons for success or failure of soil conservation projects* Soils Bulletin 64, FAO, Rome

— (1992) *Land Husbandry* Batsford, London

Hudson, N W and Cheatle R (editors) (1993) *Working with Farmers for Better Land Husbandry* IT Publications, London

Hunting Technical Services Ltd (1983) *Review of the Rural Development Areas Programme*, final report and annexes

Hurault, J (1970) 'L'organisation du terroir dans les groupements bamileke' *Etudes rurales* no 37–9, pp 232–56

Hurni, H (1986) *Guidelines for Development Agents on Soil Conservation in Ethiopia* CFSCDD, Ministry of Agriculture, Addis Ababa

Hurni, H and Kebede Tato (1992) *Erosion, Conservation and Small-Scale Farming* ISCO/Geographia Bernensia, Berne

ICRA (1991) *Farming Systems Study of the Matengo Highlands, Mbinga District, Tanzania* ICRA, Wageningen

IFAD (1990) *Upper east region land conservation and smallholder rehabilitation project*, appraisal report, International Fund for Agricultural Development, mimeo

— (1992) *Soil and water conservation in sub-Saharan Africa: Towards sustainable production by the rural poor* International Fund for Agricultural Development, Rome

IFPRI (1995) *A 2020 Vision for Food, Agriculture and the Environment* IFPRI, Washington, DC

Igbokwe, E M (1985) *Acceptance of Agricultural Innovations and Its Socio-economic Impact on Farmers in Awgu L.G.A., Nigeria* MSc thesis, Nsukka, Department of Agricultural Extension, University of Nigeria

IUCN (1990) *Ethiopian Natural Resources Conservation Strategy* IUCN, Gland

IUCN/UNEP/WWF (1991) *Caring for the Earth: A Strategy for Sustainable Living* IUCN, Gland

Kolawole, A (1988) 'Cultivation of the Floor of Lake Chad: A Response to Environmental Hazards in Eastern Borno, Nigeria' *Geographical Journal*, vol 154, no 2, pp 243–50

Lal, R (1993) 'Soil Erosion and Conservation in West Africa' in D Pimentel (editor) *World Soil Erosion and Conservation* Cambridge University Press, Cambridge

Lipton, M (1987) 'Limits to price policy for agriculture. Which way for the World Bank?' *Policy Development Review*, vol 5, pp 197–215

Lowdermilk, W (1935) 'Civilization and soil conservation' *Rhodesia Agriculture Journal* vol 32, pp 553–57

LRDC (1978) *Land Resources of Central Nigeria: Agricultural Development Possibilities* vol 2A, The Jos Plateau Land Resources Study 29, LRDC, Ministry of Overseas Development, London

Lutz, E, Pagiola, S and Reiche, C (1994) 'The costs and benefits of soil conservation: the farmers' viewpoint *The World Bank Research Observer*, vol 9, pp 273–95

McAllister, P (1992) 'Rural production, land use and development planning in Transkei: a critique of the Transkei Agricultural Development Study' *Journal of Contemporary African Studies*, vol 11, no 2, pp 200–22

McDaniel, J B (1966) 'Some government measures to improve African agriculture in Swaziland' *Geographical Journal*, vol 13, pp 506–13

McIntosh, A, Quinlan, T and Vaughan, A (1993) *Promoting Small Scale Irrigation Enterprises in the Transkei: Possibilities and Constraints* Institute for Social and Economic Research, University of Durban-Westville, Durban

McKenzie, B (1984) *Ecological Considerations of Some Past and Present Land Use Practices in Transkei*, unpublished PhD thesis, University of Cape Town, Cape Town

Malawi Government (1992) *Smallholder Agricultural Services Project Preparation Exercise Final Report*, Ministry of Agriculture, Lilongwe

Mando, A, van Driel, W F and Zombré, N P (1993) *Le role des termites dans la restauration des sols ferrugineux tropicaux encroutés au Sahel*, publication de l'Antenne Sahélienne, Université Agricole de Wageningen, Wageningen

Mansfield, J E, Bennet, G, King, R B and Lawton, R M (1976) *Land Resources of the Northern and Luapula Provinces, Zambia: A Reconnaissance Assessment*, vol 4, The Bio-Physical Environment Government Printers, Lusaka, Zambia

Meadows, D et al (1972) *The Limits to Growth. A Report for the Club of Rome's Project on the Predicament of Mankind* Earth Island, London

Millar, D (1992) *Understanding rural peoples' knowledge and its implications for intervention: "from the roots to the branches" Case studies from Northern Ghana*, MSc thesis, Agricultural University, Wageningen

Million, Alemayehu (1992) *The Effect of Traditional Ditches on Soil Erosion and Production. On-Farm Trials in Western Gojam, Dega Damot Awraja* Soil Conservation Research Project Research Report 22, Berne, Switzerland

Milne, G (1947) 'A soil reconnaissance journey through parts of Tanganyika territory, December 1935–February 1936' *Journal of Ecology*, vol 35, nos 1 and 2, pp 192–265

Ministry of Agriculture (1978) *Overgrazing and livestock development*, report of a seminar, Mbabane

Ministry of Agriculture/Shawel Consult (1988) *Study of traditional conservation practices* Ministry of Agriculture, Addis Ababa

Morgan, W T W (1979) (editor) *The Jos Plateau: a survey of environment and land use* Occasional Publications (new series) no 14, Department of Geography, University of Durham, Durham

— (1985) 'Foreword' in A A Adepetu *Farmers and their farms on four fadamas on the Jos Plateau* JPERDP Interim Report no 2, Department of Geography, University of Durham, Durham, pp i–ii

Morin, S (1988) *Colonisation agraire, dégradation des milieux et refus d'innovation dans les hautes terres de l'Ouest-Cameroun* Journées de Geographie Tropicale, Bordeaux, p 26

Mortimore, M and Tiffen, M (1995) 'Population and environment in time perspective: the Machakos story', pp 69–90 in T Binns (editor) *Population and Environment in Africa* Wiley, Chichester

Mukanda, N and Mwiinga, R (1993) 'Soil and water management in Zambia', pp 37–40 in N Hudson and R Cheatle (editors) *Working with Farmers for Better Land Husbandry* IT Publications, London

Murdoch, G (1970) *Soils and land capability in Swaziland* Ministry of Agriculture, Mbabane

Mwenda, E (1963) 'Historia na Maendeleo ya Ubena' *Swahili*, XXXIII, no 2

Myers, R J K and Foale, M A (1981) 'Row Spacing and Population Density in Grain Sorghum: A Simple Analysis' *Field Crops Research*, vol 4, pp 147–54

Nachtigal, G (1980) *Sahara and Sudan* vol II: *Kawar, Bornu, Kanem, Borku, Ennendi 1879–1889* Hurst & Co, London

National Population Commission (1992) *Census News* 3, no 1, National Population Commission, Lagos

Ngapgue, J N (1994) *Mutations des milieux agraires en pays bamileke: L'exemple des vallées à raphiales du village Bafou dans le départment de la Menoua (Ouest-Cameroun)*, geography dissertation, University of Yaounde I, p 119

Ngoufo, R (1989) *Les Monts bamboutos, environnement et utillisation de l'espace*, PhD, University of Yaounde, p 349 + 1 atlas

Norman, M J T, Pearson, C J and Searle, P G E (1984) *The Ecology of Tropical Food Crops* Cambridge University Press, Cambridge

North, R (1995) *Life on a Modern Planet. A Manifesto for Progress* Manchester University Press, Manchester

Norton, A J (1987) *Conservation Tillage: what works*, paper presented to the Natural Resources Board Workshop on Conservation Tillage, Institute for Agricultural Engineering, Harare

Oldeman, L, Hakkeling, R and Sombroek, W (1990) *World map of the status of human induced soil degradation* International Soil Reference Centre, Wageningen

Oppong, C (1973) *Growing up in Dagbon* Ghana Publishers Corporation, Tema, Ghana

Ostberg, W (1986) *The Kondoa Transformation: Coming to Grips with Soil Erosion in Central Tanzania* Research Report 76, SIAS, Uppsala

Ostrom, E (1990) *Governing the Commons. The Evolution of Institutions for Collective Action* Cambridge University Press, Cambridge

Page, S L J and Page, H E (1991) 'Western hegemony over African agriculture in Southern Rhodesia and its continuing threat to food security in independent Zimbabwe' *Agriculture and Human Values*, vol 8, pp 3–18

Painter, T (1993) 'Getting it Right: Linking Concepts and Action for Improving the Use of Natural Resources in Sahelian West Africa' *Drylands Network Programme*, Issues paper no 40. IIED, London

Phillips-Howard, K D, Adepetu A A and Kidd, A D (1990) *Aspects of change in fadama farming along the Delimi River, Jos L G C (1982–1990)* JPERDP Interim Report no 18, Department of Geography, University of Durham, Durham

Phillips-Howard, K D and Schoeneich, K (1992) *The irrigation potential of water resources on the Jos Plateau: a preliminary analysis* JPERDP Interim Report no 27

Pike, A H (1938) 'Soil conservation among the Matengo tribe' *Tanganyika Notes and Records*, 6: 79–81

Pimentel, D *et al* (1995) 'Environmental and economic costs of soil conservation and conservation benefits' *Science*, vol 267, pp 1117–123

Place, F and Hazell, P (1993) 'Productivity effects of indigenous land tenure systems in sub-Saharan Africa' *American Journal of Agricultural Economics*, vol 75, pp 10–19

Porter, G and Phillips-Howard, K D (in press) 'Farmers, labourers and the company: exploring relationships on a Transkei contract farming scheme' *Geoforum*

Pretty, J and Shah, P (1994) *Soil and water conservation in the twentieth century: a history of coercion and control*, Research Series 1, Rural History Centre, University of Reading

Pretty, J and Chambers, R (1994) 'Towards a learning paradigm: new professionalism and institutions for agriculture', pp 182–202 in I Scoones and J Thompson (editors) *Beyond Farmer First: Rural People's Knowledge, Agricultural Research and Extension Practice* IT Publications, London

Reardon *et al* (1994) 'Links between non-farm income and farm investment in African households: adding the capital market perspective' *American Journal of Agricultural Economics*, 76, 1172–1176.

Reij, C (1983) *Evolution de la lutte anti-erosive en Haute-Volta depuis l'independence: vers une plus grande participation de la population* IES, Free University, Amsterdam

— (1991) *Indigenous soil and water conservation in Africa* Gatekeeper Series 27, Sustainable Agriculture Programme, IIED, London

— (1994) 'Building on Traditions: The Improvement of Indigenous Soil and Water Conservation Techniques in the West African Sahel' in T L Napier, S M Camboni and S A El-Swaify *Adopting Conservation on the Farm: An International Perspective on the Socioeconomics of Soil and Water Conservation* Soil and Water Conservation Society, Ankeny, Iowa, USA, pp 143–56

Reij, C, Mulder, P and Begemann, L (1988) *Water Harvesting for Plant Production* World Bank Technical Paper 91, World Bank, Washington, DC

Repetto, R (1988) *Economic policy reform for natural resource conservation* Environment Department Working Paper 4, World Bank, Washington, DC

Richards, P (1985) *Indigenous Agricultural Revolution: Ecology and Food Production in West Africa* Hutchinson, London

— (1989) 'Agriculture as a Performance' in R Chambers and L-A Thrupp (editors) *Farmer First: Farmer Innovation and Agricultural Research* IT Publications, London

Roe, E (1991) 'Development narratives, or making the best of blueprint development' *World Development*, vol 19, pp 287–300

— (1995) 'Except Africa: Postscript to a special section in development narratives' *World Development*, vol 23, pp 1065–69

Roose, E (1988) *Diversité des stratégies traditionnelles et modernes de conservation de l'eau et des sols en milieu soudano-sahélienne d'Afrique occidentale; influence du milieu physique et humain* Communication ISCO VI, Addis Ababa

— (1989) 'Méthodes traditionnelles de gestion de l'eau et des sols en Afrique occidentale soudano-sahélienne; définitions, fonctionnements, limites et améliorations possibles' *Bulletin Reseau Erosion*, vol 10

Roose, E, Dugué, P and Rodriguez, L (1992) 'La G.C.E.S. Une nouvelle stratégie de lutte anti-érosive appliquée à l'aménagement de terroirs en zone soudano-sahélienne du Burkina Faso' *Revue Bois et Forêts des Tropiques*, no 233, pp 49–63

Roose, E, Kaboré, V and Guenat, C (1994) *The Zaï practice: a West African traditional rehabilitation system for semi-arid degraded lands. A case study in Burkina Faso,* paper presented at the seminar on the Rehabilitation of Degraded Land, Tunisia, November 1994

Rotenhan, D F von (1966) *Bodennutzung und Viehhaltung im Sukumaland, Tanzania: die Organisation des Landbewirtschaftung in Afrikanischen Bauernbetrieben* Afrika Studienstelle, IFO-Institut für Wirtschaftsforschung (München), Springer, Berlin

Runge-Metzger, A (1988) *Variability in agronomic practices and allocative efficiency among farm households in Northern Ghana: A case study in on-farm research* Nyankpala Agricultural Research Report, no 2, Margraf Scientific Publishers, Weikersheim

Scoones, I (1991) 'Wetlands in drylands: key resources for agricultural and pastoral production in Africa' *Ambio*, vol 20, pp 366–71

Scoones, I et al (1996) *Hazards and Opportunities. Farming Livelihoods in Dryland Africa. Lessons from Zimbabwe* Zed Books, London

Scoones, I and Thompson, J (editors) (1994) *Beyond Farmer First: Rural People's Knowledge, Agricultural Research and Extension Practice* IT Publications, London

Scott, P (1951) 'Land policy and the native population in Swaziland' *Geographic Journal* vol CXVII, part 4, pp 435–57

SCRP (1987) *Soil Conservation Progress Report* Ministry of Agriculture, Addis Ababa

Shah, P (1994) 'Participatory watershed management in India: the experience of the Aga Khan Rural Support Programme', pp 117–23 in I Scoones and J Thompson (editors) *Beyond Farmer First: Rural People's Knowledge, Agricultural Research and Extension Practice* IT Publications, London

Shaxson, T F, Hudson, N, Sanders, D, Roose, E and Moldenhauer, W (1989) *Land Husbandry: A Framework for Soil and Water Conservation* Soil and Water Conservation Society, Ankeny, IA

Showers, K (1989) 'Soil erosion in the kingdom of Lesotho: origins and colonial response, 1830s–1950s' *Journal of Southern African Studies*, vol 15, pp 263–89

Soil Conservation, Land Use and Water Programming Administration (SCLUWPA) 1986–90, Various Office Reports, Ministry of Agriculture, Khartoum

Southall, R (1982) *South Africa's Transkei: The Political Economy of an 'Independent' bantustan* Heinemann, London

Spaargaren (1977) 'Estimated soil loss due to sheet erosion', Ministry of Agriculture, Land Use Planning Division, Mbabane

Stenhouse (1944) 'Agriculture in the Matengo Highlands' *East African Agricultural Journal* vol X, July

Stocking, M (1985) 'Soil conservation policy in colonial Africa' *Agricultural History*, vol 59, pp 148–61

— (1986) *The cost of soil erosion in Zimbabwe in terms of the loss of three major nutrients* AGLS Working Paper 3, FAO, Rome

— (1993) *Soil erosion in developing countries: where geomorphology fears to tread!* School of Development Studies Discussion Paper 241, University of East Anglia, Norwich

— (1996) 'Challenging conventional wisdoms about soil erosion in Africa' in M Leach and R Mearns (editors) *The Lie of the Land: Challenging Received Wisdom in African Environmental Change and Policy* James Curry, London

Swift, J (1996) 'Desertification: narratives, winners and losers' in M Leach and R Mearns (editors) *The Lie of the Land: Challenging Received Wisdom in African Environmental Change and Policy* James Curry, London

Tanzania/Netherlands Farming Systems Research Project (1989) *Diagnostic survey of Maswa and Meatu Districts (Phase 1: Informal Survey)* Agricultural Research Institute Ukiruguru and Royal Tropical Institute Amsterdam Working Paper no 4

Tchawa, P (1991) *La dynamique des paysages sur la retombée meridionale des hauts plateaux de l'Ouest-Cameroun*, PhD, University of Bordeaux III, p 400

Thrupp, L A (1987) *Building Legitimacy of Indigenous Knowledge. Empowerment for Third World People, or Scientised Packages to be Sold by Development Agencies?* IDS, University of Sussex, Brighton

Tiffen, M, Mortimore, M and Gichuki, F (1994) *More People, Less Erosion: Environmental Recovery in Kenya* John Wiley, Chichester

Toulmin, C (1993) 'Combating Desertification: setting the agenda for a global convention' *Drylands Network Programme* Issues Paper no 42, IIED, London
— (1994) *Gestion de terroirs: concept and development* United Nations Sudano-Sahelian Office
Trapnell, C G (1953) *The Soils, Vegetation and Agriculture of North-eastern Rhodesia* Government Printer, Lusaka
Turner, B, Hyden, G and Kates, R (editors) (1993) *Population Growth and Agricultural Change in Africa* University of Florida Press, Gainesville
UN (1992) *Agenda 21: The United Nations Plan of Action from Rio* United Nations, New York
Unamma, R P A *et al* (editors) (1985) *Farming Systems in Nigeria: Report of the Bench-Mark Survey of the Farming Systems of the Eastern Agricultural Zone of Nigeria* AERLS, National Root Crops Research Institute, Umudike
UNCOD (1977) *Round up, Plan of Action and Resolutions* Conference on Desertification, 29 August–9 September 1977, United Nations, New York
UNEP (1984) *General Assessment of Progress in the Implementation of the Plan of Action to Combat Desertification, 1978–1984* Report of the Executive Director, UNEP, Nairobi
URT/EEC (1987) *Regional Agricultural Development Plan. Iringa Region. Final Report* EEC Project no 5100.33.50.007, Agrar-und Hydrotechnik GMBH, Essen
USAID (1978) Swaziland RDA Infrastructure Support Project. (Paper 645-0068) Mbabane
Veterinary and Agricultural Department (1942–1946) *Annual reports* Government Printer, Mbabane
Walling, D (1988) 'Measuring sediment yield from river basins' in R Lal (editor) *Soil Erosion Research Methods* Soil and Water Conservation Society, Angkeny, pp 39–74
Wardell, A (1991) *DANIDA/GOT Identification Mission on Soil and Water Conservation/Agroforestation in Njombe and Makete Districts, Iringa Region, Tanzania. Identification Report and Preliminary Project Proposals for Njombe and Makete Districts* DANIDA, Dar-es-Salaam
Warren, D (1991) *Using indigenous knowledge in agricultural development*, Discussion Paper 127, World Bank, Washington, DC
Whitlow, J R (1988) 'Soil Conservation History in Zimbabwe' *Journal of Soil and Water Conservation*, vol 43, no 4, pp 299–303
Williams, G J and Eades, N W (1939) *Explanation of the geology of degree. Sheet no 18 (Shinyanga)* Department of Land and Mines, Geological Division Bulletin no 13
Wilson, K (1988) *Indigenous conservation in Zimbabwe: soil erosion, land use planning and rural life*, paper presented September 1988, African Studies Association Conference, Cambridge
World Bank (1993) *Social Indicators of Development 1993* Johns Hopkins University Press, Baltimore, MD
Wright, P (1982) *La gestion des eaux de ruissellement* OXFAM/ORD du Yatenga (Burkina Faso)
Yohannes, G Michael (1989) *Land Use, Agricultural Production and Soil Conservation methods in the Andit Tid Area Shewa Region* Soil Conservation Research Project Research Report 17, Berne, Switzerland

ACRONYMS AND ABBREVIATIONS

ADD	Agricultural Development Division
ARPT	Adaptive Research Planning Team
CBDA	Chad Basin Development Authority
EHRS	Ethiopian Highlands Reclamation Study
HDS	Harmonious Development in the Sahel
IFAD	International Fund for Agricultural Development
ISWC	Indigenous Soil and Water Conservation
NGO	Non-governmental Organization
NRDEPD	Natural Resources Development and Environmental Protection Department
PRA	Participatory Rural Appraisal
RDA	Rural Development Area
RDP	Reconstruction and Development Programme
SAD	Djenné Agricultural Systems Project
SAP	Structural Adjustment Programme
SCIP	South Chad Irrigation Project
SCLUWPA	Soil Conservation, Land Use and Water Programming Administration
SCRP	Soil Conservation Research Project
SODECOTON	Société de Développement du Coton
SWC	Soil and Water Conservation
TSCP	Transkei Soil Conservation Programme
UNCOD	United Nations Conference on Desertification
UNEP	United Nations Environment Programme
VCW	Village Community Worker
VIDCO	Village Development Committee

LIST OF AUTHORS

H El Abbassi	Département de Geographic, Université de El Jadida, Morocco.
J K Adewumi	Department of Agricultural Engineering, Ahmadu Bello University, Zaria, Nigeria.
M Ait Hamza	Faculté des Lettres et Sciences Hymaines, Université Mohammed V, Rabat, Morocco.
Ben Anamoh	Northern Region Rural Integrated Programme, Tamale, Ghana.
Kebede Asrat	Christian Relief and Development Association, Addis Abeba, Ethiopia.
Roy Ayariga	Upper Region Agricultural Development Programme, Ghana.
S Bisanda	Southern Highlands Zonal Research and Training Center, Mbeya, Tanzania.
M Chaker	Faculté des Lettres et Sciences Humaines, Université Mohammed V, Rabat, Morocco.
Ousmane Cissé	CARE International, Djenné, Mali.
S M A Dabloub	Institute for Environmental Studies, University of Khartoum, Sudan.
Gouro Dicko	CARE International, Djenné, Mali.
Yaya Doumbia	CARE International Djenné, Mali.
Berhanu Fantaw	Soil Conservation Research Project, Addis Abeba, Ethiopia.
J Hagmann	Agritex/GTX Conservation Tillage Project, Masvingo, Zimbabwe.
Abdou Hassan	Programme Special National FIDA/Niger, Illéla, Niger.

245

F Hiol-Hiol	Department of Forestry, University of Dschang, Cameroon.
Kederalah Idris	Christian Relief and Development Association, Addis Abeba, Ethiopia.
E M Igbokwe	Department of Agricultural Extension, University of Nigeria, Nsukka, Nigeria.
M B Jilambu	Borno State Agricultural Development Programme, Maiduguri, Borno State, Nigeria.
Vincent Kaboré	Appui-Recherche-Coopération, Ouagadougou, Burkina Faso.
Kajela Kefeni	Soil Conservation Research Project, Addis Abeba, Ethiopia.
Armand Kassogue	Projet Hydraulique, Bandiagara, Mali.
Are Kolawole	Center for Social and Economic Research, Ahmadu Bello University, Zaria, Nigeria.
Mamadou Komota	Service des Eaux et Forêts, Bandiagara, Mali.
Hans-Joachim Krüger	Soil Conservation Research Project, Addis Abeba, Ethiopia.
A Laouina	Département de Geographie, Université Mohammed V, Rabat, Morocco.
Anderson Lema	Department of Geography, University of Dar es Salaam, Tanzania.
J H Mangisoni	Department of Rural Development, Banda College of Agriculture, Lilongwe, Malawi.
Mouga Masdewel	Antenne Sahélienne, Agricultural University of Wageningen, Ouagadougou, Burkina Faso.
A C Mbegu	HADO Project, Kondoa, Tanzania (at present Sumbawangu, Tanzania).
Yohannes G Michae	Soil Conservation Research Project, Addis Abeba, Ethiopia.
David Millar	Tamale Archdiocesan Agricultural Programme, Tamale, Ghana.
Alemayehu Million	Soil Conservation Research Project, Debre Berhan, Ethiopia.
K Murwira	ITDG Chivi Food Security Project, Chivi, Zimbabwe.
T N Mwambazi	Soil Productivity Research Programme, Kasama, Zambia.
D Ndoum Mbeyo	Department of Rural Education, University of Dschang, Cameroon.
J A Ngailo	National Soil Service, Tanga, Tanzania.
Chris Oche	Department of Geography, University of Transkei, Umtata, South Africa.
P E Odo	Department of Crop Science, University of Maiduguri, Maduguri, Nigeria.

M Osunade	Obafemi Awolowo University, Ife-Ife, Nigeria.
Matthieu Ouedraogo	Projet Agro-Forestier, Ouahigouya, Burkina Faso.
K Phillips-Howard	Department of Geography, University of Transkei, Umtata, South Africa.
G S Phiri	Blantyre Agricultural Development Division, Blantyre, Malawi.
Chris Reij	Center for Development Cooperation Services, Vrije Universiteit Amsterdam, The Netherlands.
Justin Sagara	Développement en Harmonie au Sahel, Bandiagara, Mali.
M O El Sammani	Institute for Environmental Studies, University of Khartoum, Sudan.
Boubacar Sanogo	CARE International, Djenné, Mali.
Ferdinand Schutgens	Projet Hydraulique, Bandiagara, Mali.
Ian Scoones	Institute of Development Studies, University of Sussex, Brighton, UK.
Mesfin Semegn	Ministry of Agriculture, Eastern Harerge Office, Ethiopia.
J M Shaka	National Soil Service, Tanga, Tanzania.
Patrick M Sikana	Adaptive Research Planning Team, Kasama, Zambia.
Maja Slingerland	Antenne Sahélienne, Agricultural University of Wageningen, Ouagadougou, Burkina Faso.
F Tchala Abina	Department of Rural Education, University of Dschang, Cameroon.
Paul Tchawa	Department of Geography, University of Yaounde, Cameroon.
A E M Temu	Southern Highlands Zonal Research and Training Centre, Mbeya, Tanzania.
Camilla Toulmin	International Institute for Environment and Development, London, UK.
Joanna Wedurn	CARE International, Djenné, Mali.
J M Wickama	National Soil Service, Tanga, Tanzania.
A M Yagoub	Institute for Environmental Studies, University of Khartoum, Suden.
M I Yakamba	Borno State Agricultural Development Programme, Maiduguri, Borno State, Nigeria.

INDEX